CW01467726

ISBN 978-1-332-22782-2
PIBN 10301198

THE GATE OF THE ROUSILLON

H. Belloc, del

THE PYRENEES

BY

H. BELLOC

WITH FORTY-SIX SKETCHES
BY THE AUTHOR
AND TWENTY-TWO MAPS

METHUEN & CO.
36 ESSEX STREET W.C.
LONDON

900355

First Published in 1909

PREFACE

THE only object of this book is to provide, for those who desire to do as I have done in the Pyrenees, a general knowledge of the mountains in which they propose to travel.

I have paid particular attention to make clear those things which I myself only learned slowly during several journeys and after much reading, and which I would like to have been told before I first set out. I could not pretend within the limits of this book, or with such an object in view, to write anything in the nature of a Guide, and indeed there are plenty of books of that sort from which one can learn most that is necessary to ordinary travel upon the frontier of France and Spain ; but I proposed when I began these few pages to set down what a man might not find in such books : as—what he should expect in certain inns, by what track he might best see certain districts, what difficulties he was to expect upon the crest of the mountains, how long a time crossings apparently short might take him, what the least kit was which he could carry into the hills, how he had best camp and find his way and the rest, what maps were at his disposal, the advantages of each map, its defects, and so forth. The little of general matter which I have admitted into my pages—a dissertation upon the physical nature of the chain, and a shorter division upon its political character—I have strictly limited to what I thought necessary to that general understanding of a mountain without which travel upon it would be a poor pleasure indeed.

If I have admitted such petty details as the times of trains, and the cost of a journey from London, it is because I have found those petty details to be of the first importance to myself, as indeed they must be to all those who have but little leisure. I have in everything attempted to set down only that which would be really useful to a man on foot or driving in that country, and only that which he could not easily obtain in other books. Thus I have carefully set down directions as minute as possible for finding particular crossings and camping grounds, for the finding of which the ordinary Guide Book is of no service. My chief regret is that the book will necessarily be too bulky to carry in the pocket; for it is meant to be not so much a lively as an accurate companion to the general exploration of those high hills which have given me so much delight.

CONTENTS

LIST OF ILLUSTRATIONS

LIST OF MAPS

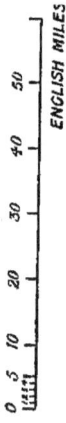

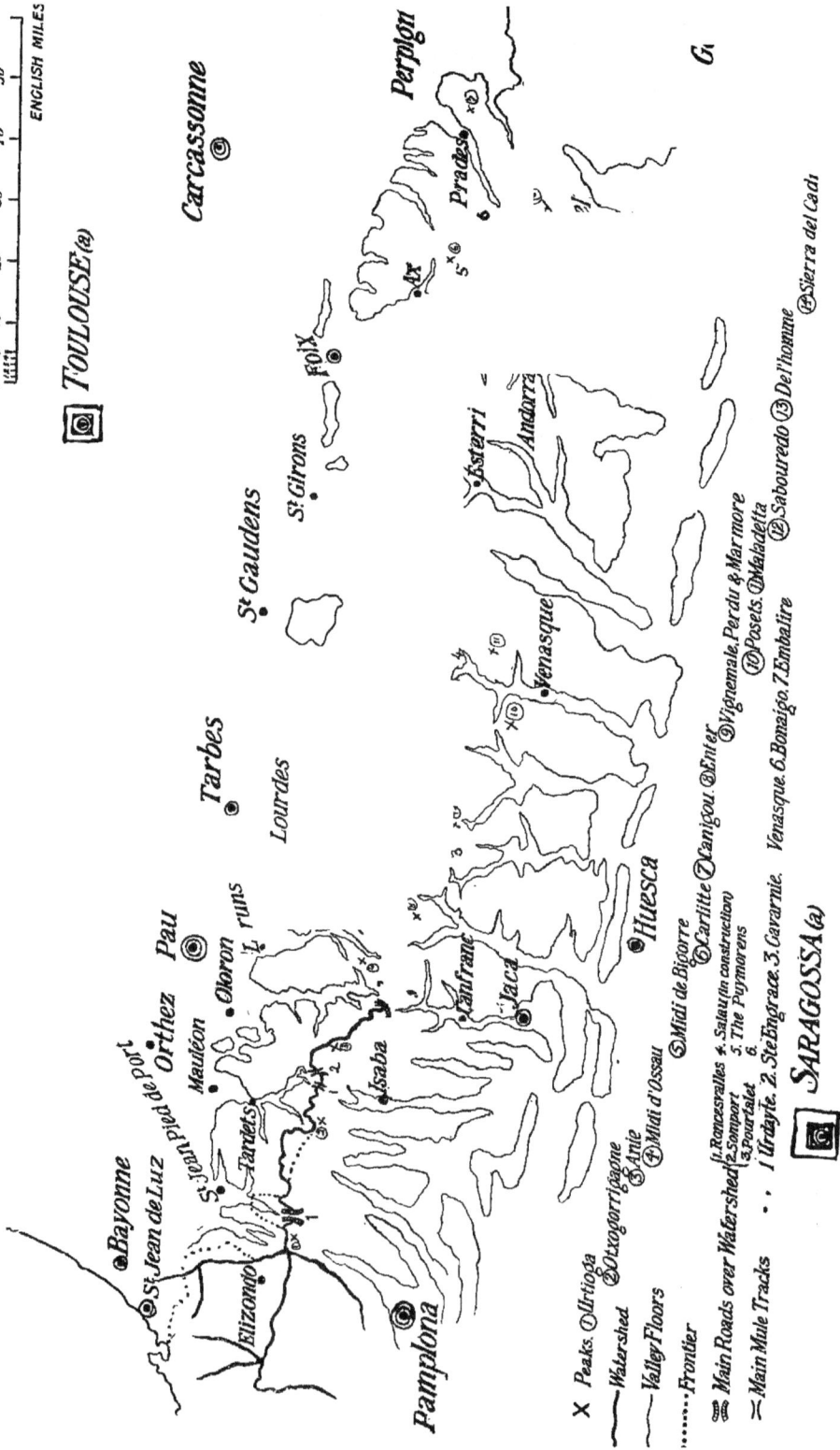

ENGLISH MILES

0 5 10 20 30 40 50

TOULOUSE (a)

Carcassonne

Perpign

G

Prades

Ax

5

Foix

St Girons

Esterri

Andorra

St Gaudens

Venasque

Tarbes

Lourdes

Huesca

Pau

Orthez

Oloron

Mauléon

ruins

Bayonne

St Jean de Luz

St Jean pied de Port

Tardets

Elizondo

Pamplona

Anie

Isaba

Canfranc

Jaca

SARAGOSSA (a)

X Peaks. ①Artioga
— Watershed ②Orxogorrigagne
— Valley Floors ③Anie
.... Frontier ④Midi d'Ossau
⑤Midi de Bigorre
⑥Carlitte ⑦Canigou ⑧Enter
⑨Vignemale. Perdu & Marmore
⑩Posets. ⑪Maladetta
⑫Sabouredo ⑬Del'homme
⑭Sierra del Cadi

🛤 Main Roads over Watershed {1.Roncesvalles 4.Salau(in construction)
2.Somport 5. The Puymorens
3.Pourtalet 6.
•• Main Mule Tracks 1 Urdayte. 2. SteEngrace.3.Gavarnie. Venasque. 6.Bonaigo.7.Embalire

GENERAL

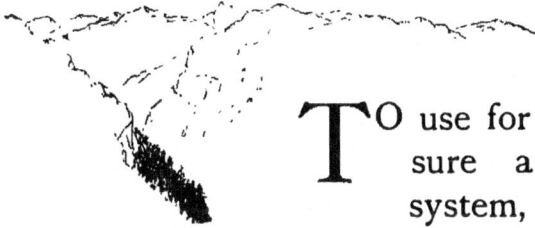

TO use for travel or for pleasure a great mountain system, the first thing necessary is to understand its structure and its plan ; to this understanding must next be added an understanding of its appearance, climate, soil, and, as it were, habits, all of which lend it a character peculiar to itself.

These two approaches to the comprehension of a mountain system may be called the approaches to its physical nature ; and when one has the elements of that nature clearly seized, one is the better able to comprehend the human incidents attached to it.

From an appreciation of this physical basis one must next proceed to a general view of the history of the district—if it has a history—and of the modern political character resulting from it. At the root of this will be found the original groups or communities which have remained unchanged in Western Europe throughout all recorded time. These groups are sometimes distinguishable by language, more often by character. Changes of philosophy profoundly affect them ; changes of economic circumstance, though affecting them far less, do something to render the problem of their continuity complex : but upon an acquaintance with the living men concerned, it is always possible to distinguish where the boundaries of a country-side are set ; and the permanence of such limits in European life is the chief lesson a deep knowledge of any district conveys.

The recorded history of the inhabitants lends to these hills their only full meaning for the human being that visits them to-day ; nor does anyone know, nor half know, any countryside of Europe unless he possesses not only its physical appearance and its present habitation, but the elements of its past.

These things established, one can turn to the details of travel and explain the communications, the difficulties, and the opportunities attaching to various lines of travel. In the case of a mountain range, the greater part of this last will, of course, for modern Englishmen, consist in some account of wilder travel upon foot, and the sense of exploration and of discovery which the district affords.

Such are the lines to be followed in this book, and, first, I will begin by laying down the plan and contours of the Pyrenees.

The first impression reached by modern and educated men when they consider a mountain system is one over-simple. This over-simplicity is the necessary result of our present forms of elementary education, and has been well put by some financial vulgarian or other (with the intention of praise) when he called it "Thinking in Maps," or, "Thinking Imperially"; for the maps in a man's head when he first approaches a new range are the maps of the schoolroom.

Thus one sees the Sierra Nevada in California as one line, the Cascade Range as another parallel to it. The Alps and the Himalayas alike arrange themselves into simple curves, arcs of a circle with a great river for the cord. The Atlas is a straight line cutting off the northern projection of Africa, the Apennines are a straight line running down the centre of Italy. Such are the first geographic elements present in the mind.

The next impression, however, the impression gathered in actual travel, or in a detailed study, is one of mere confusion, a confusion the more hopeless on account of the false simplicity of the original premise. Deductions from that premise are perpetually at variance with the observed facts of travel or of study, the exceptions become so numerous as to swamp the rule, and an original misconception upon the main character of the chain prevents a new and more accurate synthesis of its general aspect. Thus, the conception of the Cascade Range upon the Pacific Coast of the United States as parallel and separate from the Sierras, confuses one's view of all the district round Shasta, and of all the watersheds south of the Mohave where the two systems merge; or again, one who has only thought of the Alps as a mere arc of a circle misconceives, and is bewildered by the nature, the appearance, and the whole history of the great re-entrant angles of the Val d' Aosta with its Gallic influences ; the anomaly of the Adige Valley will not permit him to explain its political fortunes, and the outlying arms which have preserved the independence of Swiss institutions upon the southern slope will not fall into his view of the mountains.

This confusion, I say, is not due so much to the multitude of detail as to the permanent effect of an original strong and over simple conception remaining in the mind as it continues to accumulate increasing but sporadic knowledge of a particular district ; and it is a confusion in which those who have formed such an erroneous conception commonly remain.

In order to avoid such confusion and to allow one's increasing knowledge a frame wherein to fit, it is essential to grasp in one scheme the few elementary lines which underlie a mountain system, and such a scheme will be a trifle more complex than the too simple scheme usually presented, but once one has it one can appreciate the place of every irregularity in the structure of the whole chain of hills.

In the case of the Pyrenees the common error of too great simplicity may be easily stated. These mountains are regarded as a wall separating France from Spain, and running direct from sea to sea. Such an aspect of the range will more and more confuse the traveller and reader the more he studies the actual shape of the valleys. Another picture should occupy the mind, and it will presently be seen that with this picture permanently fixed as a framework for the whole system, an increased knowledge of its details does but expand the sense of unity originally conveyed. The Pyrenees must not be regarded as a sharp heaped ridge forming a single watershed between the plains of Gaul and those of Northern Spain, and running east and west from the Atlantic to the Mediterranean. They form a system, the watershed of which does not exactly stretch from one sea to the other. The axis of it does not consist of one line ; the general direction is not due east. The axis of the Pyrenæan chain is built up of two main lines, of approximately equal length : the one running south of east from a point at some distance from the Atlantic, the other north of west from a point right on the shores of the Mediterranean. These two lines do not meet. They miss by over eight miles, and the gap between them is joined by a low saddle.

The first of these lines starts from a point (Mont Urtioga), 25 miles south of the corner made by the Bay of Biscay at Irun, and some fifteen miles west of its meridian ; it runs about 9° 15′ south of east to the peak called Sabouredo, the last of the Maladetta group, the direct distance from which to Mount Urtioga is precisely 200 kilometres or 124 miles. The second runs from Cape Cerberus on the Mediterranean to a peak called the Pic-de-l'homme, which stands a trifle over 12 kilometres, or 7¾ miles north of the Sabouredo ; its direction is 9°.25′ north of west, and the total length of this second line is just over 190 kilometres, or 117 miles.

The simplest scheme, then, in which we can

regard the Pyrenees, is as a system of not quite parallel lines of equal length, running one towards the other, but missing by not quite 8 miles; the gap or "fault" joined by a zigzag saddle on the watershed. The westernmost of these lines splits into several branches before it reaches the Atlantic, so that the true western end of the chain lies well to the south and east of that ocean (at Mount Urtioga); the other starts from, and forms a projection in, the Mediterranean. The full distance as the crow flies from Mount Urtioga to Cape Cerberus upon the Mediterranean is 390 kilometres, that is 241 miles. And there is but 10 kilometres, or 6½ miles difference in length between the two halves of the chain.

If U be the point called Mount Urtioga, S the Sabouredo, L the Pic-de-l'homme, and C Cape

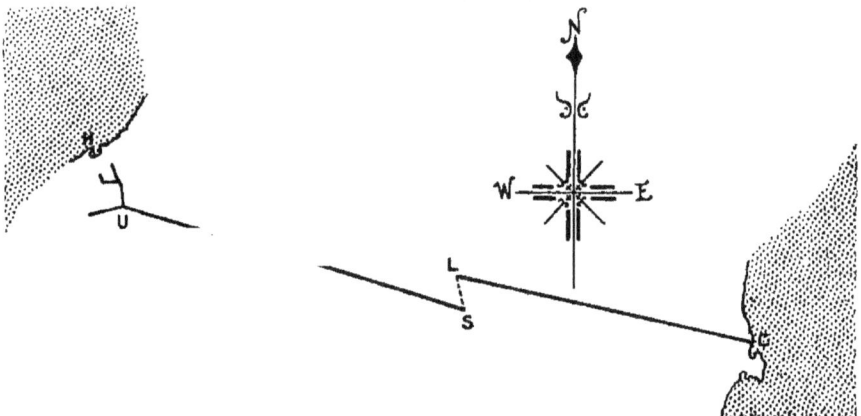

PLAN

Cerberus, these two lines and the gap between them will lie precisely as in this plan.

With this main guide by which to judge the structure of the chain, all details will be found to fit

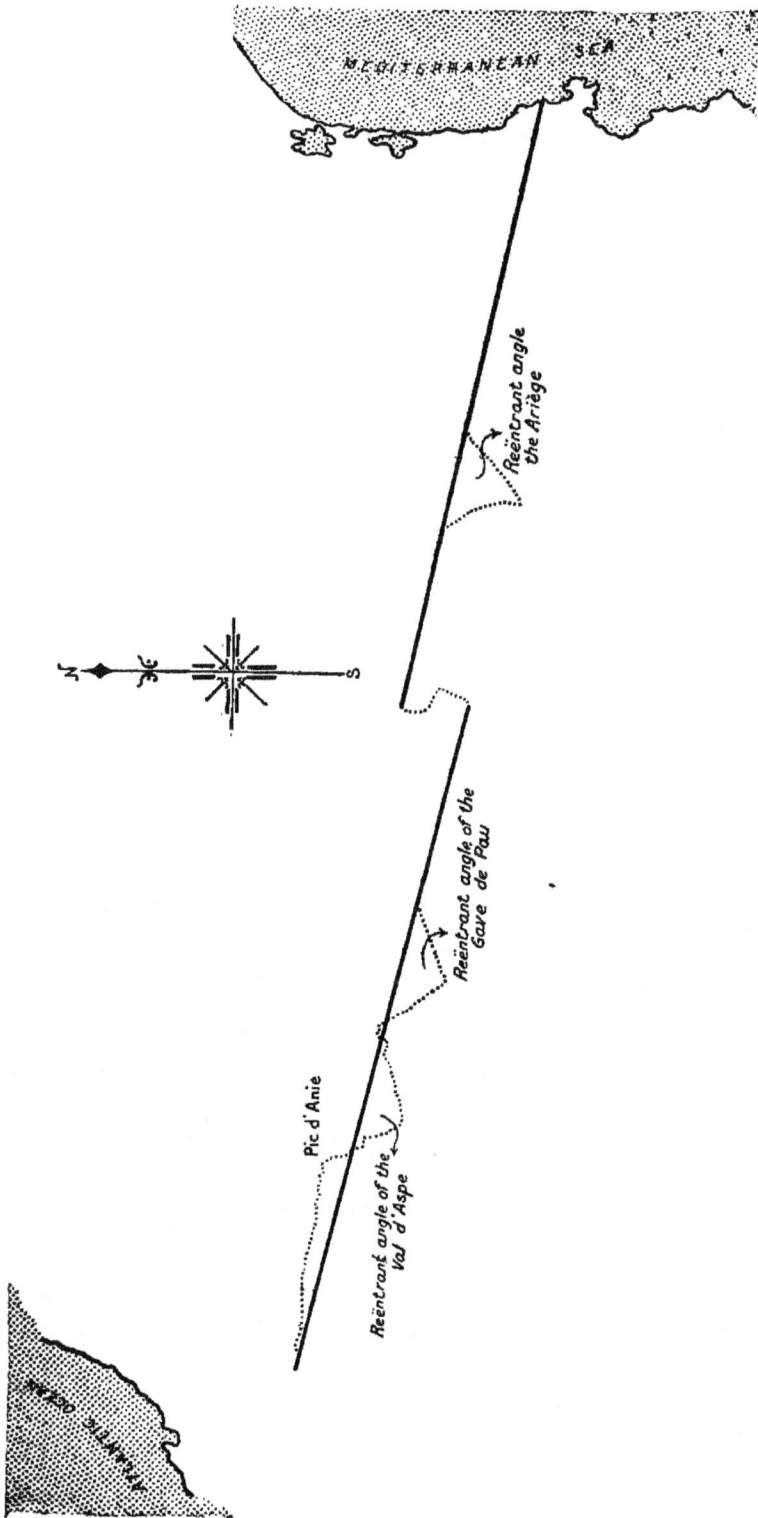

PLAN B

in, and the two first variations which we must
superimpose upon so general a view, are to be
found in the "step" or "corner" formed by the
watershed round the Pic d'Anie. The southward
turn of the range is here not gradual but sharp, and
the Somport, the pass at the head of the Val d'Aspe,
lies almost a day's going below the Port St Engrace,
which is the Pass near the Pic d'Anie. Next, one
should note the two re-entrant angles, one to the
north of the chain, one to the south, which distinguish the Spanish valley of the Gallego and the
French valley of the Gave de Pau respectively.
These features modify the simplicity of the first or
western branch of the chain ; one exceptional feature
only modifies the second or Eastern branch, and this
is the deep re-entrant wedge of the Ariège valley
upon the French side. We may therefore regard
the elements of the watershed somewhat according
to the sketch plan B on the preceding page.

The details of the watershed when they are
given in full are of course indefinitely more numerous
and complicated, and it may be of advantage for
those who would understand the structure of the
Pyrenees to glance also at the plan opposite, where the
dotted line represents the exact trace of the watershed,
the dark lines the simple structure described above.

The watershed then should be regarded as the
chief feature in the range, and as the backbone of the
whole system. Geologically, it is not the foundation
of the range. Geologically, the range was piled up
by the junction of a number of short separate ranges,
each of which ran with a sharper south-eastern dip

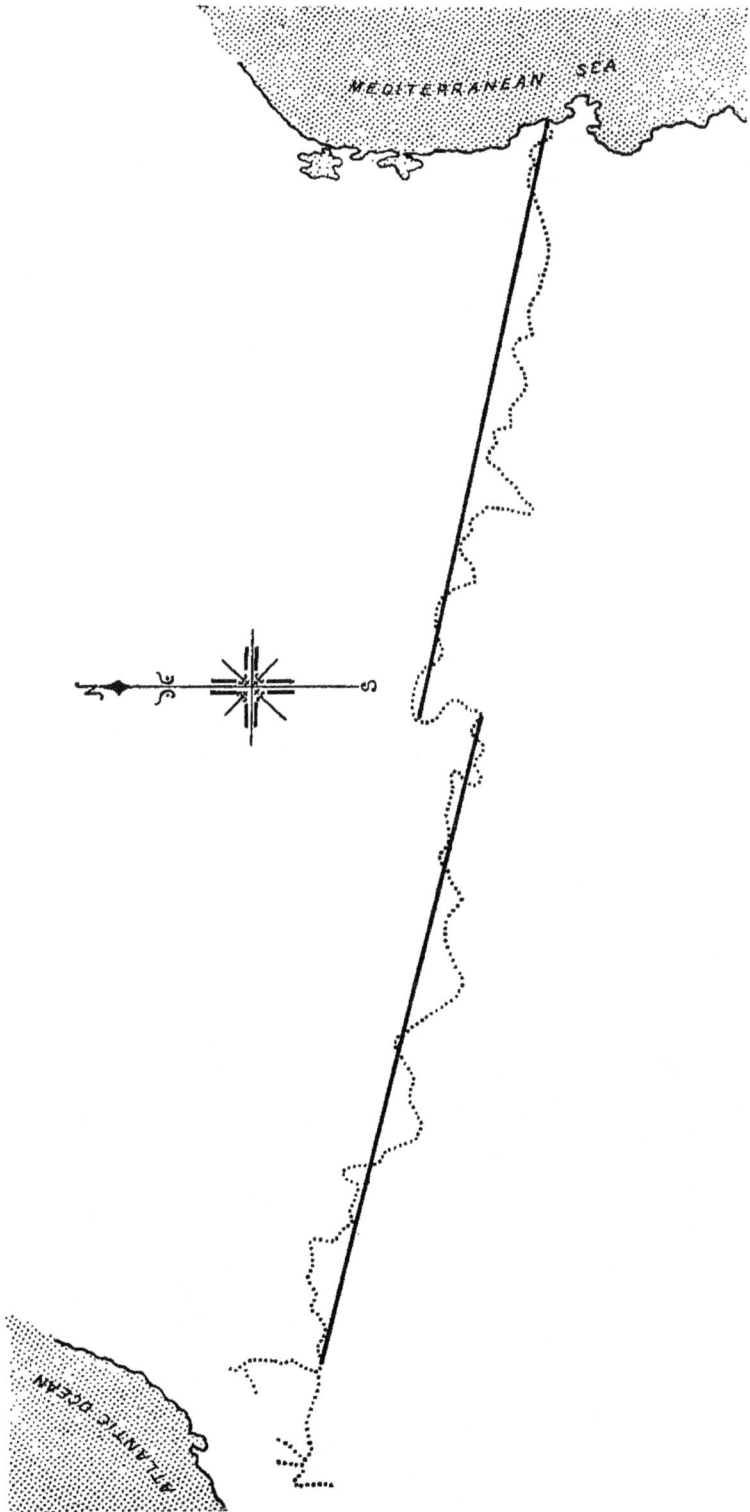

MEDITERRANEAN SEA

ATLANTIC OCEAN

PLAN C

(about 30°) than does the present long line of saddles which has joined them and forms the existing water-shed, and probably the process of the formation of the Pyrenees was upon the model sketched in the following diagram.

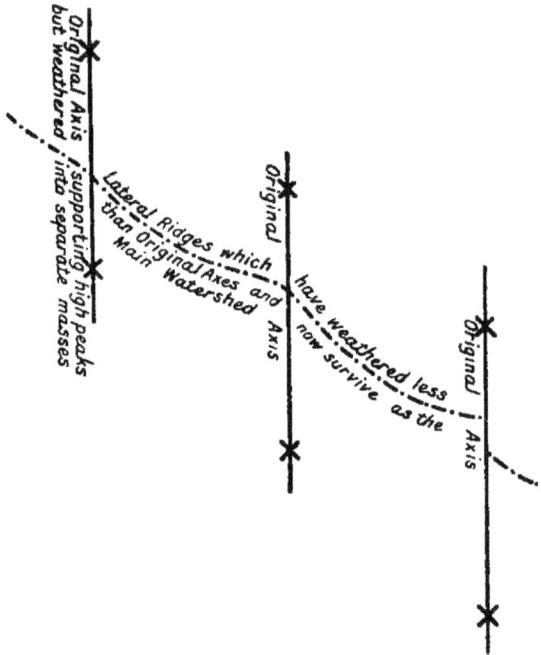

PLAN D

But for the purpose of understanding the Pyrenees as they now are, it is the existing watershed which we must consider, and that runs as I have said.

Next, the rule should be laid down that the Pyrenees must be separately considered on their northern and upon their southern slopes. It will be seen later that the physical and historical contrast between the two sides of the mountains is sometimes acute and sometimes slight, but the contrast between the general contour upon either side is such as to make it impossible to unite both in one similar system.

The Northern slope of the Pyrenees is narrow and precipitous. The plains are for the greater part of its length clearly separated from the mountains; the easy country in some places (at St Girons, for instance, and in the Flats between Lourdes and Tarbes) is not 20 miles as the crow flies from the highest peaks.

On the Spanish side, on the contrary, the mountainous district will run from two to three times that distance. Its extreme width between the open country at the foot of the Sierra Monsech and the Salau Pass is over 60 miles, and it is nowhere less than two days good journey on foot from the summits to the plains.

This differentiation between the northern and the southern slopes is not merely one of width, it is due to profound differences in the contours which makes the Spanish side of the system a different type of mountain group from the French. For, on the French side the Pyrenees consist in a series of great ribs or buttresses running up from the plains perpendicularly to the main heights of the range, and it is between these ribs or buttresses that the separate and highly distinct valleys which are the characteristic habitations of the French Basques and Bearnais lie. On the Spanish side the main structure is in folds *parallel* to the watershed; the lateral valleys descending from the watershed run southward for but a very short distance, they come, within a few miles, upon high east-and-west ridges which sometimes rival the main range itself in height and which succeed each other like waves down to the plains of the Ebro. The contrast in structure north and south of the watershed may be expressed in the formula of this plan. A man

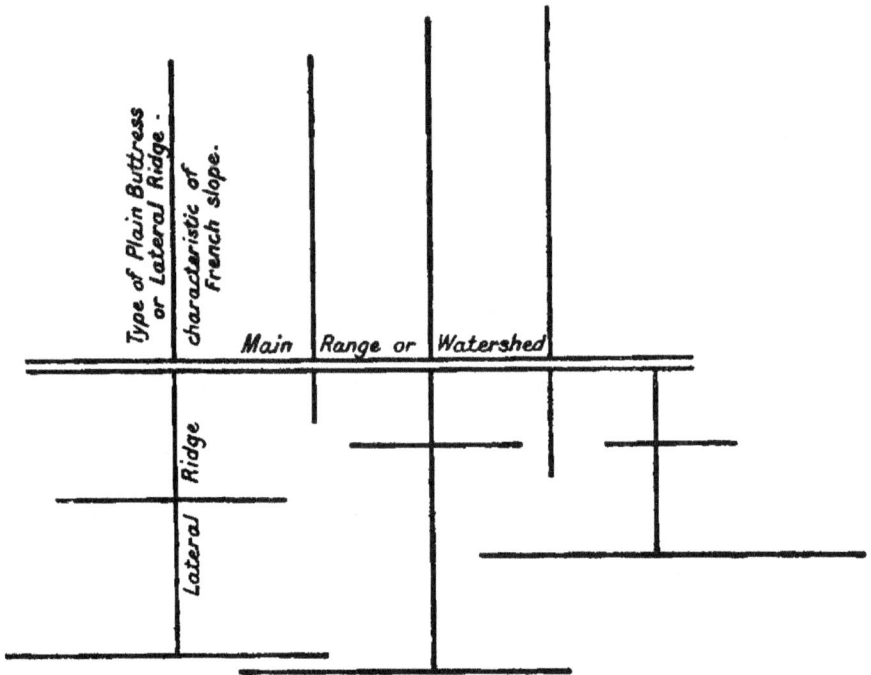

Type of Secondary Lateral Ridges characteristic of Spanish slope.
These just miss making a continuous line and through the narrow gaps between them, the rivers they deflect reach the Ebro.

PLAN E

looking at the Pyrenees from the French towns at their base sees in one complete view a belt of steep rising slopes, and a long fairly even line of summits against the sky. A man looking at the range from the Spanish plains can only in a few rare places so much as catch sight of the main range. In far the greater number of such views he will have before him a high ridge which masques the country beyond. If, then, the reader or the traveller regards the French slope as being essentially a series of profound valleys parallel to each other and running north and south, he will have grasped the main aspect of this side of the range. If he will regard the Spanish slope as a series of parallel outliers which begin quite close to the watershed, and which, though falling at last into the plains of the Ebro, are, even the most southern of them, of considerable height, he will have grasped the structure of the Pyrenees upon the side which looks towards the sun.

To these two main aspects the reader must again admit considerable modifications, the first of which concerns the French side.

This Northern slope and its valleys, of which the sketch map over page indicates the general arrangement, may be divided into two sections : the first, a western section, the second an eastern one, and these two are separated at A——B by a division roughly corresponding to the "fault" between the two axes, which, as we have seen, determine the lie of the range.

From the first, or western axis, descend with a regular parallelism, eight valleys. Each valley bifurcates in its higher part into two main ravines, and a whole system of minor streams, spread over an indefinite number of tortuous dales and gullies, attach to each valley.

There is a mark or limit for each of these western French valleys, which is the spot where

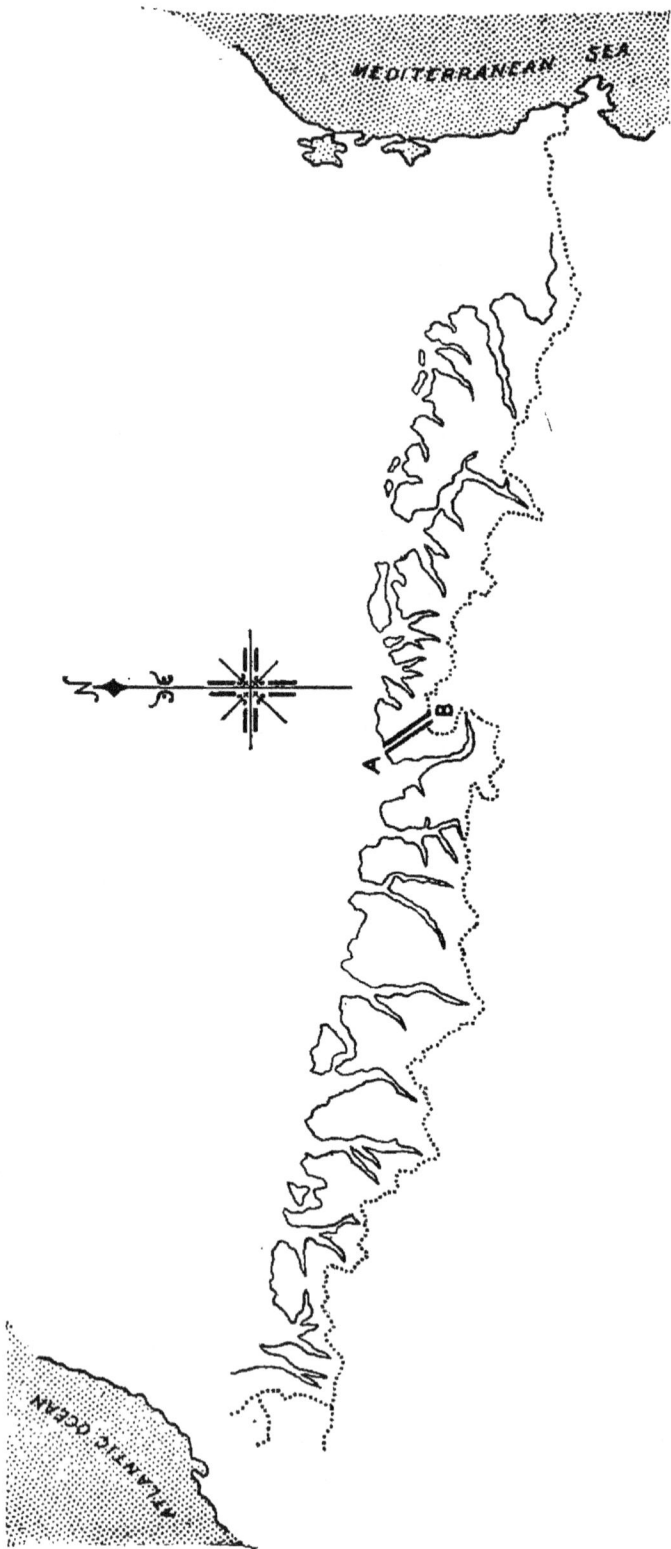

PLAN F

it debouches upon the open country. Thus the
Gave d'Ossau falls into the Gave d'Aspe at
Oloron, nevertheless the two valleys must be regarded as separate, because the meeting of the
two streams takes place in the open plain. On the
other hand the valley of Baigorry, and the neighbouring valley of St Jean, though containing two large
separate streams, must be treated as one system,
because these streams meet at Eyharee near Ossès,
and the open plain is not reached before a point
some miles further down beyond Canbo.

The test, though it may sound arbitrary upon
paper, is quite easily appreciated in the landscape,
and the separate valleys are more clearly marked,
perhaps, than those of any other European mountain
chain.

These eight valleys (see plan G over page), going
from west to east, are first that of the Nive (the
bifurcations of which gives St Jean and the Baigorry),
next that of the Gave-de-Mauleòn (Larreau and Ste
Engrace), and both of these are Basque. Next
comes the valley of the Gave d'Aspe (with the
bifurcation of Lourdios and Urdos), up which went
the main Roman road into Spain and which is the
first of the Bearnese valleys. Next is the Val
d'Ossau (with the bifurcation of Gabas and the lac
d'Arrius), next the valley of the Gave de Pau (with
the bifurcations of Cauterets and Gavarnie), next the
valley of Bigorre, a short valley bifurcating in two
minor streams at its head. Next, or seventh, comes
the Val d'Aure, with Vielle upon its western bifurcation and Bordères upon its eastern ; and lastly the

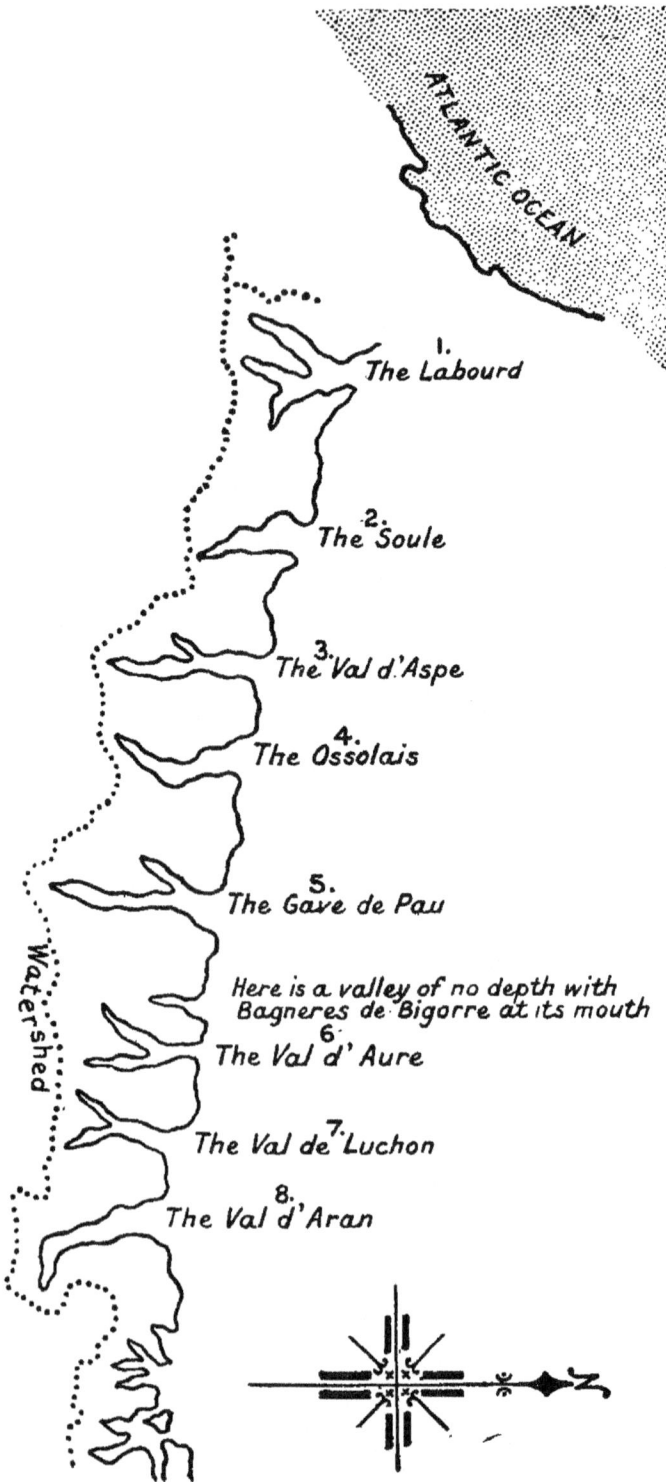

ATLANTIC OCEAN

1. The Labourd

2. The Soule

3. The Val d'Aspe

4. The Ossolais

5. The Gave de Pau

Here is a valley of no depth with Bagneres de Bigorre at its mouth

6. The Val d'Aure

7. The Val de Luchon

8. The Val d'Aran

Watershed

PLAN G

bifurcated valley of the Garonne, whose level and deep floor comes nearest of all to the main chain, and holds on the west Luchon, on a branch called the Pique, on the east Viella in the Val d'Aran.

Once past this point, the structure of the hills along the eastern run of the broken Pyrenean axis changes. The mountains here are penetrated by

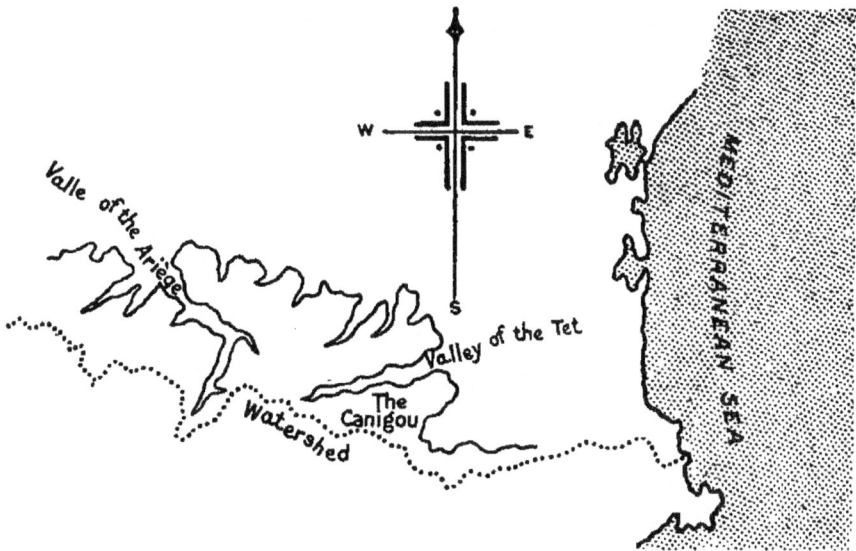

PLAN H

only two valleys, but each is much longer and more important than any of the eight just mentioned, and these two great valleys run, not parallel to each other, nor north and south as do the eight western ones, but at a steep slant : the one (that of the Ariège) goes westward, and the other (that of the Tet) eastward. Save for these two main valleys no regular features can be discovered in the eastern portion ; all is here a labyrinth of dividing and subdividing lateral ridges, and the only thing giving unity to the group is this system of two great trenches which run up towards each other, the one from the Plain of Toulouse, the other from that of Perpignan, to meet on the high land of the Carlitte group. Strictly speaking, the western valley is not wholly that of the Ariège, but those of the Ariège and Oriège combined, and it is further remarkable that no regular passage exists from the one depression to the other, but by a curious topographical accident, which will be described later in the book, the crossing from the Ariège to the Tet has to be made by going over on to the south side of the range, and then back again on to the north side.

The importance of these two main valleys upon the eastern half of the northern slope of the Pyrenees is sufficiently evident from the historical fact that each determines a great historical district : the one, that of the Ariège, was the country of *Foix*, the other that of the Tet, was the *Rousillon*. And while the eight small western valleys running parallel to each other separate local customs and dialect alone, the ridge of the Ariège and the Tet may almost be said to have separated two nationalities, and owed ultimate allegiance for a thousand years, the one to a Gallic, the other to an Iberian lord.

Beyond the valley of the Tet and eastward of the Canigou runs the little fag end of the range, which falls into the sea at Cape Cerberus, and is called the "Alberes." Here there is but little distinction between the northern and the southern side, the general shape of a sharp ridge is maintained throughout, but the height lowers more and more as the sea is approached. These hills are everywhere passable ; the ancient road into Spain which crosses them, should count, geographically and historically, rather as a road crossing round the Pyrenees at their sea end, than as a road crossing the chain.

A Pyrenean valley upon the French side always presents the same main characteristic, and this is true not only of the main valleys, but of the innumerable lateral valleys which ramify from the main valleys in all directions.

The characteristic of these French Pyrenean valleys is that they are sharply divided by very narrow gorges into two or more level basins. These level basins in the smaller valleys and on the high levels where there is pasturage and no habitation are called " Jasses "; the large and low ones are called " Plains " or " Plans "; but they are the same in their essential feature, which is a level floor more or less wide, bounded by the steep hills upon either side, and ending and beginning with a rocky gate through which the valley stream cascades. The whole formation suggests the former existence of great and small lakes, which burst their way through the gorges at some remote time.

These gorges are very rarely of any length, a point in which the Pyrenees differ from the Alps. Here and there, especially in the limestone formations, you do get long and difficult passages. One, the Cacouette in the Western Pyrenees, in the upper waters of the Gave-de-

Mauléon, is not only very profound but absolutely impassable, like the Black Cañon of Colorado, but it does not lead from one part of a valley to another. It occupies the whole of the upper valley; and in general, you will not find a Pyrenean stream running, as do the Alpine streams, for some miles between precipices.

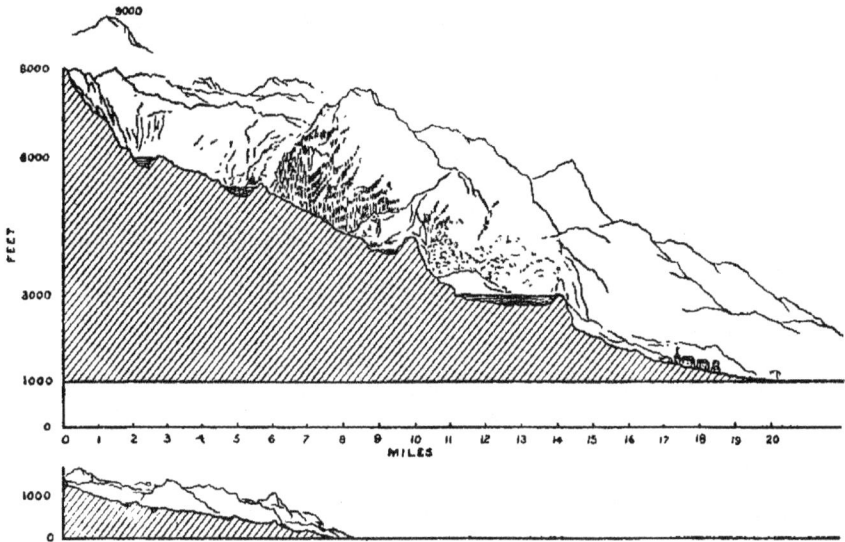

PLAN I

Each main valley has a clearly marked mouth where it debouches upon the plain; by this I do not mean that perfectly flat land comes up to and meets the hills in every case; on the contrary, at the mouth of most of these valleys are moraines left by old glaciers, but I mean that the character and aspect of the hills visibly and immediately changes, and that each of the valleys has a distinct final " gate" where it meets the lowlands, just as a river will meet the sea at a definite mouth. Now each of these openings has its characteristic town.

Mauléon, for instance, is at the mouth of the last Basque valley, Oloron at the mouth of the Val d'Aspe, Lourdes at the mouth of the valley of Argelès, etc.

Further, these towns at the mouths of the valleys have invariably chosen for their site, whether they be prehistoric or medieval, some rock on which to build a citadel; and in every case a castle is still to be found holding that rock. Lourdes, Foix, Mauléon are excellent examples of this.

Higher up the valley, the first plain above the mouth will, as a rule, contain the first mountain town. Thus Argelès lies above Lourdes, Bedous and Accous above Oloron, Laruns in the first flat of the Val d'Ossau, etc.

According to the length of the valley and the number and size of the Jasses, there may be one or more such towns enclosed by the mountain sides; thus in the valley of Lourdes we have Argelès, and above it Luz; in the valley of Soule we have Tardets above Mauléon, and higher · still we have Licq. But all the valleys, whether they contain one or more of these upland towns, have, just under the last watershed, a hamlet or village usually giving its name to the Port or Col—that is *the Pass into Spain*—above it, and the reason of this is evident enough; habitations were necessary as a place of departure and arrival for the crossing of the mountains. Of such are Gavarnie, Urdos, Merens, and the rest. These high villages have least history, least wealth, and until recently had the worst communications. For much the greater part of the year they are lost in snow, and there was an interval between the making of the great roads and the beginning of modern tourist travel when they were in peril of destruction. The new great roads drew away wealth and visitors from all but a very few, and but for the beginning of modern mountaineering they had hopelessly decayed. Even so famous a place as Gavarnie, the best known of all the valley heads was dying in the middle of the century. There are days now when it is at the other extreme : fine days in August when, for the crowd of rich people, you might be at Tring or at some reception of the late Whittaker Wright's. Even to-day, one or two of them, however clean or kindly, are odd in the way of poverty. I have known one where they had no butter and never had had any butter, and another where I was charged 8d. instead of 5d. for a bed because it was the season.

The typical Spanish valley differs, in the centre of the Chain at least, from the typical French valley. With the exception of Andorra (which reminds one in all features of the French side ; for it has the same enclosed plain, the same steps and rocky gorges between, the same Jasses, and the same arrangement of towns and villages) the greater part of the valleys, whether Catalan or Aragonese, are not only broader and their streams larger than on the French side, but their arrangement also is different, most of them lack wide pasturages, nearly all of them lack enclosed plains, and there has been no motive to penetrate them since the building of the new roads, for travel upon this side is rare.

The Spanish valley, therefore, often many days' walking in length, never direct, and forming a sort of little province to itself, will have towns and illages scattered in it, haphazard and thinly. Very often a considerable town will be found at the very end of the valley, as Esterri in that of the Noguera Pallaresa, or Venasque in that of the Esera. The lateral communications from one Spanish valley to the next are usually more difficult than those between the French valleys; for many months they are impossible, and there is no such general arrangement of towns on the plain holding the approaches to the valleys as in France, for the reason that the whole plan of the mountains on the Spanish side is far more troubled and irregular.

Thus the first town of the Aragon is Jaca; but Jaca is right in the mountains, and nothing at the outlet of the hills 50 or 60 miles down the valley makes a head town for Jaca. Jaca is a bishopric on its own. On the Gallego there is nothing but a succession of villages of which Sallent right up at the head of the valley is among the largest: it is almost a little town and so is Biescas close by. The Cinca and the Esera have indeed a town upon the plains at Barbastro, but the Noguera Ribagorzana has none, nor has its sister the Noguera Pallaresa, while the Segre has its bishopric and chief town right up in the highest hills at Urgel, and there is nothing to compare with that town until you get to Balaguer.

The southern side of the watershed differs greatly

in general structure from the northern, and must be separately recorded.

There are indeed certain accidental similarities. The enclosed valley of Andorra to the south recalls the enclosèd valley of Bedous or Accous to the north, and the very high first miles of the torrents, just under the main range, do not differ much whether they are found on the north or on the south side of the mountain. But the general plan and contour of the range presents a great contrast on either side. The main feature of the southern slope is, as I have said, a series of parallel ranges pushing out like ramparts in front of the main heights. If you follow a French valley (on the western part of the Pyrenees at least) you will finding it running fairly north and south to the point where it debouches upon the plain some 20, or 25 miles at most, from the watershed.

A Spanish valley will at first appear to have the same character, but just when you think you are in sight of the plains (for instance, just after leaving Canfranc upon the banks of the Upper Aragon) you see—beyond the first lines of flat country, and barring the view like a great wall—another high range : in this case the Sierra de la Peña, the ridge of rock which takes its name from the " Peña-de-Oroel," a mountain with its eastern end just above Jaca. Beyond this again you have the San Domingo ridge, and to the east of it, another running also east and west, the Sierra de Guara.

Pamplona again is situated at the mouth of a true Pyrenean valley (that of the Arga), not very different from the valleys to the north. It stands also on a plain, but immediately in front of it runs another range of hills, and if you climb these, you find yet another, strictly parallel and straight, standing before you and masking the approach to the Ebro. This formation in parallel outliers continues as far east as the Segre valley, that is for full three-quarters of the length of the Spanish Pyrenees, and in a sense it continues even further east than the river Segre; for the Sierra-del-Cadi, though it joins on to the main ridge at one point, is essentially an outlier in slope and formation. This parallel formation sometimes comes quite close to the central range, as for instance, in the Colorado peaks close to Sallent and Panticosa, and the long ridge to the south of Bielsa and El Plan. Indeed the characteristics of Sobrarbe, as this countryside is called, consist in these long parallel ridges.

One result of this formation is, as I have said above, that the river valleys do not run straight, as they do to the north of the range, but are thrust round at right angles when they come up against these ridges. Sometimes they will eat their way through a ridge, as do the two Nogueras, and the Arga itself south of Pamplona; but the greater part of the rivers on the Spanish side suffer the diversion of which I speak, and none more than the river Aragon, which gives its name to the whole central kingdom; for the Aragon, after having run south and straight for a few miles, like any northern river, suddenly turns westward, and runs under the foot to Sierra-de-la-Peña for two days' march. According to its first direction, it should fall into the Ebro somewhere near Saragossa; as a fact it does not come in until far above Tudela.

Another result of the formation is that the mountain tangle stretches much further on the Spanish side than it does upon the French. If you stand upon the Pass of Salau where the French have made, and the Spanish are making, a high road, you have before you to the north, at a distance of less than 10 miles, the railway and a fairly open valley. Fifteen miles at the most, as the crow flies, you have the main line and the true lowlands at St Girons. But if you turn and look out in the opposite direction over the valley of Esterri and the higher Noguera Pallaresa, you are looking over 60 miles of mountain land. From the high ridge, which is your standpoint, to the summit of El-Monsech, which is the final rampart of the hills beyond the plains of Lerida, is more than 50 miles, and the slopes of that rampart take you nearly another ten.

A further consequence of this formation is that communications are very difficult to the south of the Pyrenees. The traveller naturally ascribes the lack of communications to the character of Spanish government. It is not wholly due to a moral, but partly to a material cause. The main Spanish railway from Saragossa to Barcelona may be compared to the main French railway from Toulouse to Bayonne, but the Spanish side everywhere suffers from its great wide stretch of wild mountain land. Toulouse itself is little more than 50 miles from the crest of the mountain. Saragossa is half as much

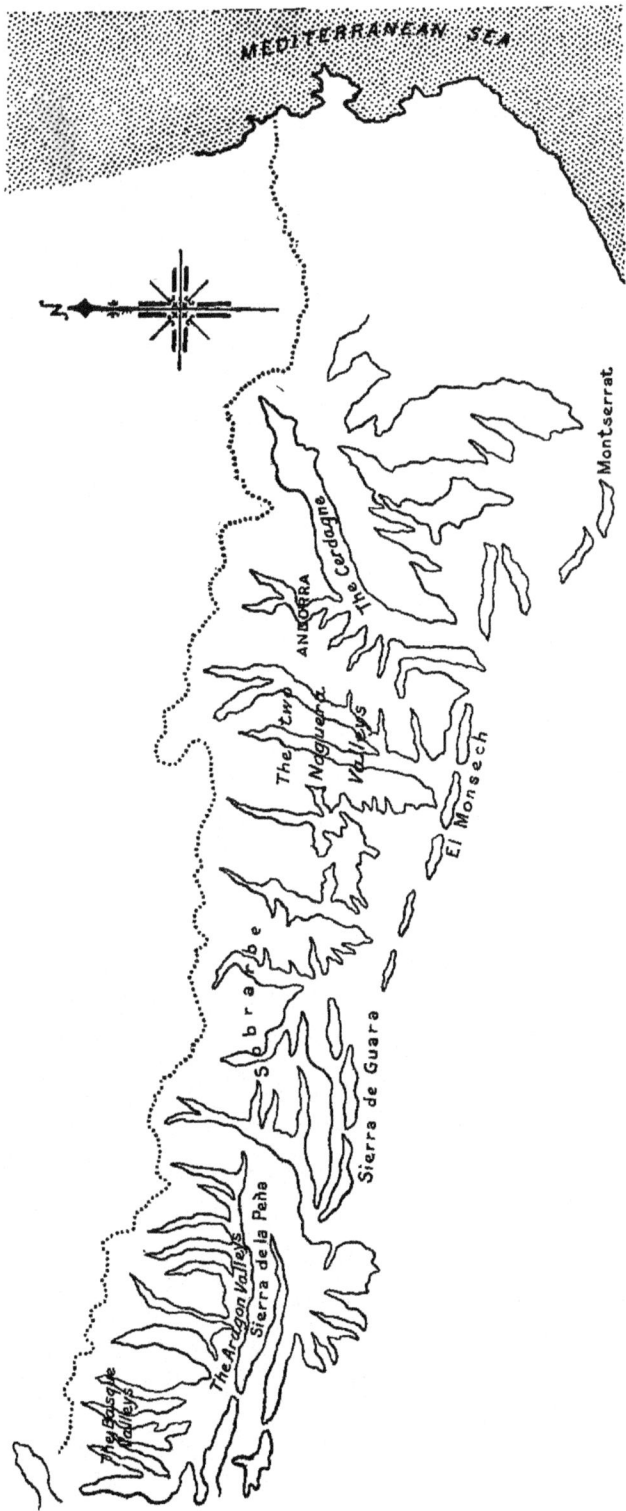

MEDITERRANEAN SEA

Montserrat

The Cerdagne

ANDORRA

The two Noguera Valleys

El Monsech

Sobrarbe

Sierra de Guara

Sierra de la Peña

The Aragon Valleys

The Basque Valleys

PLAN J

again. The Spanish Pyrenees push out civilisation, as it were, far from them. Lerida, a large town of the plains, is quite 60 miles from the watershed in a straight line. Pau or Tarbes are less than thirty. The difficulty and expense with which the civilisation of the plains, and the things belonging to it must reach the remote upper Spanish valleys largely account for the curiously high degree of their isolation from the world. Many thousands of men are born and die in those high valleys, without ever seeing a wheeled vehicle, and without knowing the gravest news of the outer world for two or three days after the towns have known it.

It is not easy in such a system to establish general divisions. We saw that this was simple enough upon the French side : eight main valleys to the west of the " fault," and two large sloping ones on the eastern limb. In the Spanish Pyrenees, the nearest thing one can get to a classification is *first* to group together the Basque valleys of Navarre, the streams of which all flow down to meet at last near Lumbier and fall into the Aragon a few miles further south. *Next* to take the group of valleys along the mouths of which stands the great Sierra de la Peña, of which the chief is the ravine of the Upper Aragon. These dales, which have at their extremities the huge masses of the Garganta and the Pic D'Anie, form the original stuff of Aragon. These few square miles were the seat from whence that race proceeded which fought its way down to the Ebro, and to the sources of the Tagus, and which can claim the Cid Campeador for its historic type. *Next* comes the group of valleys beginning with the Gallego and ending with that of Venasque, which forms the eastern limb of Aragon, and has borne for many centuries the title of " Sobrarbe." *Next* to consider the two Nogueras as the Western, the Cerdague and Andorra as the Eastern Catalonian land.

It should be noted that the fine tenacity of Spain in general, and of these hills in particular, has preserved with exactitude the ancient and natural divisions of the land. The long unbroken ridge which encloses the Basque valleys is also the frontier of Navarre. The unity of Aragon survives in the present administrative division of Jaca. The eastern valleys are still called the " Sobrarbe," and the " fault," or break between the two main lines of the Pyrenees, still forms an historical and racial break to the South as to the North of the chain. Beyond it eastward begins the Catalan language, and the next group to consider are the great Catalan valleys of the two Nogueras and of the Segre. The two Nogueras ultimately fall into the Segre, but in the mountain regions the three form three large parallel valleys, each with a character and nourishment of its own and all Catalan. Of these three, the Segre is the most striking ; its upper waters are the centre of the flat valley of the Cerdagne, the only natural passage from the North into Spain, and one of its earliest tributaries nourishes the Republic of Andorra.

East of the Segre valley, and of the Sierra del Cadi which bounds it, no classification is possible. It is a labyrinth of little valleys. A flat welter of hills running down everywhere to the sea, and narrowing at the extreme end into Cape Cerberus : these last crests, as I have said, take the name of " Alberes."

This contrast in structure between the northern and the southern side of the range runs through many other aspects of the hills beyond structure alone. We have seen that it affects the type of civilisation, leaving the deep but short French valleys far more open to the culture and influences of the plains than were the Spanish just over the watershed. There is much more.

The fall of the light is in itself a contrast. The slopes of the Spanish mountains, and especially of the high mountains, look right at the blazing sun. They are more bare of wood, much, than are the French slopes. They are more burnt. Water is less plentiful. Insects are more numerous, and there is less cultivation ; but one cannot say that there is, as a rule, a scantier population. Small villages and hamlets are rarer in the remote gorges, but small towns a little lower down are common, and apart from the population economically dependent upon summer tourists in France, it might be doubted whether the Spanish side were not as well garnished as the French ; one might venture to imagine that in the Dark, and early Middle, Ages, when the full effect of the natural condition of the mountains could be felt, the population of either side was sensibly the same.

The highest peaks are upon the Spanish side, but it does not show splendid isolated masses of rock like the two Pics-du-Midi, or lonely masses like the Canigou. On the other hand, the general character of the rocks is more savage and more fantastic, and it is upon the south side of the range that one most feels creeping over one that sentiment of unreality or of a spell, which so many travellers in the Pyrenees have been curious to note. The local names express it upon every side. There is " The Mouth of Hell," " The Accursed Mountain," " The Lost Mountain," " The Peak of Hell," " The Enchanted Hills " or " Encantados," and hundreds of other legendary titles that express, as well as do the mountain tunes, the sense of an unquiet mystery.

The Spanish side is again remarkable for true rivers running in considerable valleys everywhere east of Navarre. Though the rainfall is less upon the southern side than upon the northern, yet, because the catchment areas are broader, the streams running at the bottom of the Spanish valleys are larger and more important. A glance at the map will show upon the French side a whole series of parallel river valleys running down from the summit into the plains to join the Adour or the Gironde. Armagnac and Bearn are crowded with them. A man going eastward from Bayonne to Pau, from Pau to Tarbes, from Tarbes to Toulouse, will cross more than forty streams all spreading out like a fan from the central axis of the Lannemezan Plain; a man going eastward below the Spanish foot hills from Melida, let us say through Huesca to Lerida, will find but half-a-dozen

of such water crossings.   Again, you have between
the Soule and the Labourd, between the Val d'Aspe
and the Val d'Ossau, between the Val d'Aure and
the Garonne, distances of 8, 10, or at the most
12 miles, but the distance between neighbouring
Spanish valleys (if we except the near approach of
the Gallego and the Aragon), is always much
greater.   Between the Noguera and the Segre,
for instance, there is at the nearest point, 20
miles ; between the two Nogueras, another 20 ;
between the last of these two rivers and the Esera,
quite 20 more.   The whole Spanish side with the
exception of Navarre is thus built up of considerable valleys, comparable to those of the Tet and
the Ariege alone upon the northern slope.

The details of Pyrenean structure will concern a
man travelling on foot more than do these general
lines, though the whole aspect of the range must be
grasped before one can understand its details.   The
separate peaks and valleys, the intimate structure of
the range is remarkable everywhere for its abrupt-
ness.   It is this physical feature in the Pyrenees,
coupled with the absence of snow, which gives them
the highly individual character they bear.

Fantastic outlines are not to be discovered in
these hills so frequently as in the Dolomites, nor
are wholly isolated hills common, though such few
as exist are very striking, but for day after day a
man wandering in the Pyrenees sees cliffs more
regularly high, a greater succession of rocks more
precipitous, and a more permanent succession of

THE UPPER AND THE LOWER SLOPE

connected summits above him than in any other European range.

The absence of snow is a further sharp characteristic in the range. The essential feature of an Alpine landscape is the snow ; and it is not only the essential feature of that landscape to the eye, it is the condition which controls the lives of those who inhabit, and of those who visit, the valleys. You can still wander a trifle in Switzerland. Even to-day I have come to villages where foreigners were still thought comic, and an ignorance of the German tongue was still thought amazing. But though you can wander, your wandering is strictly limited. Above a certain line you can go forward only with technical knowledge and in a special way. You are upon ice or snow. All climbing in the Alps depends upon this, and most of travel as well. A man may pass many days for instance in the upper valley of the Rhone, and then pass many days more in the upper valley of the Aar, but to go from one to the other he must take one of two strictly defined paths, unless he is willing to undertake special work requiring technical knowledge and particular aids. The hills between the two valleys are not a field for his exploration ; they are a great mass of impassable and unapproachable land, all frozen, and diversified only by very narrow valleys inhabitable nowhere but at their base. No one could lose himself for many days upon the Wetterhorn, the Jungfrau, or the Finsteraarhorn ; you can approach these mountains for the glory of the thing, but they are not a countryside. Now in the Pyrenees almost all the surface of the mountains, say 250 miles by 60, is at your disposal. It is this and a local custom of live and let live which make the pleasure of them inexhaustible; and which, combined with certain protective methods of their own, make it certain that the Pyrenees will never be overcome or changed by men. They are too large.

This surface of some 15,000 square miles is diversified in a manner fairly continuous throughout the chain. The valley floors are given up to cultivation in their lower part, their upper parts consist of damp close pastures, and between the two types of level are to be found, as we shall presently see, sharp gates of rock through which the river saws its way in a gorge. Above the valley floor, and at the end of it, where the stream springs out below the watershed (such springs bubbling up suddenly through the porous rock are called Jeous), steep banks of two, three and four thousand feet, broken almost invariably here and there into precipices, forbid the way; and these, in perhaps half their extent, are covered with enormous woods of beech below (mixed often with oak) and of pine above.

When you have climbed up these slopes through the forest or over the naked rock, you come, in the last heights, either to large grassy spaces, which often sweep right over the summit of a lateral ridge, and sometimes extend over both sides, even, of the main watershed (as, between the Val d'Aran and Esterri) or else—more commonly—upon a jumble of jagged rocks and smooth, perpendicular or overhanging slabs, which defend the final secrets of the range.

The succession of these features is nearly universal. The only places where they are modified are the two lower ends of the range. There the rocks sink, the hills are rounded, the precipices disappear in the last Basque valleys, while the Alberes at the other extremity of the chain against the Mediterranean are at last mere toothed rocks. All between (with the exception of the Cerdagne, which is a country to itself) is built up in successive bands of valley floor, steep forest, or steep rock, broken with limestone precipices, and finally on the highest ridge sweeps of grass or jagged edges of stone.

It is this character in the last ridge of the Pyrenees that determines the nature of a passage over them, and since a passage from valley to valley is the chief business of the day when one is exploring the range, I will next describe these crossings, for the method of them is very different from that of other mountains, and has largely determined the history and customs of their inhabitants.

In other high mountains you will either find snow above a certain level and covering for most of the year most of the passes, some of the passes for all the year, or, as you go further south, you will commonly find many gaps which long years of weathering have reduced to easy slopes, or you will find great differences in slope between the one and the other side of the range; as, for instance, the difference between the long valleys that lead up eastward into the Californian Sierras, and the sharp escarpment which falls eastward upon the desert side of those heights. In all other ranges that I have seen or read of, save the Pyrenees, there is at least great diversity in the opportunities of crossing, whether natural or artificial There is great diversity, as a rule, in the natural crossings ; some are quite easy ascents and descents on either side (as the Brenner Pass over the Alps); some, though difficult, are notably lower than the average height of the range (as the Mont Genèvre from the Durance into Piedmont) ; some, these more rare, are deep gorges cleft right through the range (as the Danube gorge through the Carpathians).

Now it is characteristic of the Pyrenees that in the main part of their length no such diversities appear, save that there are two kinds of summit surfaces on the high cols, rock and grass : the grass the rarer.

If anyone looks closely at the Somport, especially noting the line which the old track took before the modern road was made, he will agree that it is a pass which, though steep, had no "edge" to it, so to speak. The grass would take any kind of traffic. The same is true of course of the Cerdagne, the only broad valley across the Pyrenees. But the Somport is well to the west end of the range ; the Cerdagne is well to the east end. All the main part between could take no vehicle, and has crossings of a kind which I shall presently describe : sharp, the escalade difficult, the first descent upon the far side, or the last ascent upon the near side, steep.

There is perhaps an exception to be found in the case of the Bonaigo, but this pass also presents difficulties to wheels upon its western side, and in the lower valley at the gorge.

In general the crossings of the Pyrenees everywhere display certain characters rare or absent in other ranges, which are *first* that they are very numerous (a feature due to the absence of snow), *secondly*, that they are very high, *thirdly*, that they hardly ever involve any true climbing, and *fourthly*, that they nearly always involve some considerable care on the part of the wayfarer and are somewhere dangerous either upon the northern or the southern side.

This can be well illustrated by a particular example in the few miles between the Pic D'Anéu and the Canal Roya. Here there is a range no part of which descends much below 2100 metres nor rises much above 2300. There are two distinct saddles where a man can cross on foot, and neither is appreciably lower than the peaks of the range, which are but lumps of rock a little higher than the grassy ridges from which they spring. Any man knowing the country and with a fairly good head could trust himself to half-a-dozen places westward of the two which I have mentioned (which are called the Col D'Anéu and the Port of Peyreget). Nevertheless the easiest of them, the Port de Puymaret, easy as it is upon the French side, gives some pause upon the Spanish. The traveller finds himself, once over the crest, within a few yards of a rocky edge, beyond which there is apparently nothing but air, and, thousands of feet beyond, the precipices of the Negras. If he will approach that rocky edge he will see that everything below it is easily negotiable, and when he has once reached the floor of the Spanish valley beneath he will perhaps wonder why it seemed so difficult from above. In truth it is not really difficult at all, but the scramble looks dangerous, and it is one which most men, other than regular climbers, would think twice about when they first saw it from above. If all this is true of the Peyreget, it is still more true of the other crossings in its neighbourhood to the right and to the left.

Were the Pyrenees surmountable at comparatively few passages, these would have been so thought out and perhaps improved as to make them regular and well-known passes, which the traveller could easily deal with. It is the very number of the crossings which add to their difficulties. The people who live upon either side are indifferent in their choice among so many difficult passages, and with the exception of one or two quite modern made roads with which I shall presently deal, there are some hundreds of Cols and Ports all having in common a character of difficulty, and few naturally so much more easy than their neighbours as to concentrate travel upon them.

This feature may be summed up in the expression that the crest of the Pyrenees is rather one long ridge slightly serrated than (as in the case of most other ranges) a succession of high mountain groups separated by low saddles.

Of all the accidents that strike one in connection with the crossing of these hills nothing strikes one more than the accident of time. A Port is always a day and a long day. Here and there quite exceptionally there may be food and shelter upon either side within six or seven hours one from the other; but as a rule if you propose to sleep under cover upon either side, your effort will demand a long summer's day, and it is best to look forward to a night camp upon the further side of the range.

Before continuing the description of these passages, or any rules by which one should be guided in attempting them, it may be well to speak for a moment of the few practised and conventional tracks.

First of these come, of course, the high roads. At present, over the frontier, these are but four in number (for the low passes to the east of the Canigou may be neglected), Roncesvalles, the Somport, the Pourtalet, and the Cerdagne. Of these the Pourtalet has been but recently opened, and is still in process of being widened upon the French side. Moreover, it is so nearly neighbouring to the Somport (there is but 8 miles between them), that it hardly affords a true alternative crossing. A fifth highroad across the watershed is that which crosses it at Porté from the valley of the Ariège into the Cerdagne, but this road is essentially a lateral one. It lies wholly in French territory, it joins the French road through the Cerdagne, and you cannot go by it down the valley of the Segre. It only crosses the watershed on account of an accidental divergence of this to the south, in the upper valley of the Ariège.

These four carriage roads are all that lead, at present, over the political boundary of the Pyrenees. Another is in construction over the Port of Salau, but it is not finished upon the Spanish side. The French desire several others to go over by the Macadou, Gavarnie, etc., but their own preparations are not completed and the Spanish are not even begun.

Apart, however, from these highroads, which are carefully graded, possess an excellent surface, and are traversable by any vehicle, there are a certain number of crossings which travel has rendered familiar, and whose facility is well known. Thus, the Embalire from the Hospitalet on the upper Ariège into the Upper Segre in Andorra is a perfectly easy slope of grass, though high. Again, the Bonaigo, though there have been natural difficulties in the lower valley to be surmounted, and though there is not even a track across it, is a perfectly easy roll of grassy land barely 6000 feet high. A high road leads as far as Esterri on the Spanish side ; another goes from France on the northern side, right up the valley of the Garonne, beyond Viella, to the paths at the very foot of the pass, so that the gap between the two highways is but a few miles in length.

The Port de Venasque again, though but a mule track, is constantly used, and, though steep and high (close upon 8000 feet), presents no difficulties at all, and is almost a highway between the two countries.

The Port de Gavarnie is similarly constantly used and may be taken like any other mountain path. Certain other passes form an intermediate category. They present no difficulties to one who is acquainted with the neighbourhood, but either the whole path is difficult to trace or its last and highest portion is dangerous, or there are precipices upon its lower slope, or in one way and another they cannot be regarded as constant and regular communications of international travel, though the inhabitants use them continually. Of such a kind is the Port d'Ourdayte ; of such a kind are the passages from the Aston into Andorra, and of such a kind are most of the passages just west of the Port de Venasque. If one applied the test of asking where the Pyrenees could be crossed in doubtful weather, not half a dozen places could be found beyond the four high roads ; and even if one were to ask in what spots they could certainly be crossed by a stranger without chance of failure, the number of passages would prove less than a score. All the rest of the ridge from the Sierra del Cadi to the Basque mountains is the rocky wall I have described, with innumerable notches more or less practicable, but all difficult, and nearly all requiring a detailed knowledge of either slope.

There are one or two other features needing explanation before I close this introduction to a physical knowledge of the range ; thus the reader should be acquainted with the many groups of lakes and tarns which stand just under the highest peaks and ridges in groups : they are highly characteristic of the Pyrenees. There is a cluster of half a dozen at the western base of the Pic du Midi d'Ossau, another cluster surrounding the neighbourhood of Panticosa, another in the summit of the Encantados between the Maladetta group and the valley of the Noguera, another very famous one well known to fishermen high up in the knot of mountains whose summit is the Carlitte, and there are many isolated small lakes which the map discovers. But whether in groups or isolated, one feature is common to all these lakes of the Pyrenees—first, that none is of any size ; secondly, that all, or very nearly all, are quite in the highest parts of the hills immediately under the last escarpment ; and thirdly, as a consequence, that it is rare to find a lake which the presence of wood and the neighbourhood of habitation render suitable for camping.

It is worth remembering that, unlike most mountain systems, the Pyrenees do not, even in sudden storms, endanger one as a rule by a rapid increase in size of the torrents ; one has not to fear spates so much as one might imagine from the multiplicity of the streams on the northern side or the large area of the valleys on the southern. This truth, of course, must not be exaggerated nor too much advantage taken of it. That part of a stream which will be just traversible after several fine days may become just too violent to cross after a few hours of rain, but I have never seen those sudden changes of level from a rivulet to a considerable torrent which one may so often see in British mountains, which are common enough in Scandinavia and even in the Alps, and which are a regular condition of travel in the Rockies.

Why this should be so it would be difficult to say. The great area of forest upon the north might account for regularity upon that slope, but it would not account for it upon the Spanish side. And one would imagine that snow in large masses, which is lacking in the Pyrenees and present in the Alps, would rather tend to regulate the flow of rivers; but whatever be the cause, the evenness of level is what one used to other ranges will first remark when he has to cross and recross under different conditions the higher streams of this chain in summer.

There should lastly be noted the absence of any important glaciers, a feature due to the absence of snow fields. On the summit of the Cirque de Gavarnie, on the summits of the Pic d'Enfer and the neighbourhood, on the summits of the Maladetta group, and in one or two other parts, there are small glaciers, but they form no general feature in the landscape of the Pyrenees, and have no effect upon travel.

Lastly, the climate of these mountains should be noted: it is a very important part of the conditions which determine travel upon them.

The rain-bearing winds blow from the Atlantic eastward, and if the Pyrenees stood upon either slope equally accessible to the sea, it is possible that the Spanish side would be the more deeply wooded

and the best watered.   The sudden trend westward
of the Spanish coast, however, at the corner of the
Bay of Biscay, causes the wet winds from the Ocean
to lose most of their moisture to Galicia and the
Asturias, before they can strike the Pyrenees themselves from the south, while the same winds,
coming around the range from the north, come upon
the Pyrenees immediately after leaving the sea.
The result of this is that the French side is throughout its length more heavily watered than the Spanish
side ; but on either side there are three zones which,
though not sharply distinguishable one from the
other, are sufficiently remarkable.

The first is that of heavy rains, and, what is more
important for purposes of travel, of continuous rain
and frequent mist.   It stretches all along the western
end of the range, and only begins perceptibly to
change with the heights of the Pic d'Anie and the
precipitous barrier of the upper valley of the Aspe.
West of this line—that is, in all the Basque-speaking
country—you have deep pastures upon either side of
the range, and all the marks of the damp in the
timber and the mode of building, the vegetable
growth and the animals of the place.   Snow falls
later here than in the other parts of the Pyrenees,
for the double reason that the neighbourhood of the
sea makes the climate milder and that the hills
are less high.   In most places, for instance,
communication is not cut off between the north
and south valleys of the Basques, and men can
usually cross from Ste Engrace to Isaba at all
seasons.

The next zone (the eastern frontier of which is very vague) may be said to stretch, according to the year and the accident of weather, certainly as far as the Catalans and the valley of the Noguera on the east, and sometimes as far as the valley of the Segre itself. In all this central part of the range (which may normally be said to include more than half its length) the French or northern side is densely wooded and heavily watered, the Spanish side more dry and bare ; but even the French side slowly shows a change of climate as one goes eastward, the forests remain as dense, the rivers as full, but the days are certainly finer and mist less frequent. On the Spanish side the change as one goes eastward is less striking, because the whole climate is drier. It is to be remarked that if mist gathers upon the northern side of the hills when one is attempting a pass, one may fairly count upon its disappearance upon the Spanish side in this section; and, in general, the whole of the southern slope, from the valley of the Aragon to that of the Noguera, is of a dry and equal nature, somewhat barren and burnt, not only from the lack of moisture, but also from exposure to the sun.

The lack of moisture on the central Spanish slope, by the way, is not a little aided by the curious formation of the frontier of Navarre, and the separation between the Basques and the Aragonese; this consists in a long ridge of high land, the upper part of which is known as the Sierra Longa, which runs south and a little west from the Pic d'Anie. The effect of this lack

of moisture and excess of heat upon the central
Spanish side is not only felt on the heights of the
mountain, but also and more particularly when
one approaches the Plains. These in France are
northern in type, full of greenery, and amply watered.
In Spain, on the contrary, they are quite arid, and
if one comes in to Huesca by train upon a September evening, and looks out the next morning
over the flats that run up to the Sierra de Guara,
one has all the impressions of a desert, though
these lands are heavy corn-bearing lands in the
summer.

Finally, the third, or eastern, section of the hills
is Mediterranean in character throughout. The
Canigou is much more heavily watered than the
Sierra del Cadi, its corresponding Spanish height.
But the olives on the lower slopes, the carpet of
vineyards on the flats, the presence everywhere of
bright insects, the quality of the light and the aridity
of everything which does not happen to be planted
with trees, gives to this eastern corner of the
Pyrenees the same aspect that you may notice on
the Mediterranean hills of Southern France, Liguria,
or Algeria or the Balearic Islands, for all these
landscapes are of one kind, and binding them all
together is not only their burnt red look, but also
that tideless intense blue of the Mediterranean, the
hot white towns, and everywhere the Lateen sail
upon the coasts.

These differences of climate also determine the
seasons in which the mountains may best be visited,
for the Basque district is at your service (especially

in its western part), from the spring to the late autumn of the year; the central valleys can be everywhere travelled in only from late June to mid-September; the eastern end, again, from the Segre and beyond it, is open to you from spring to autumn.

THE POLITICAL CHARACTER OF THE PYRENEES

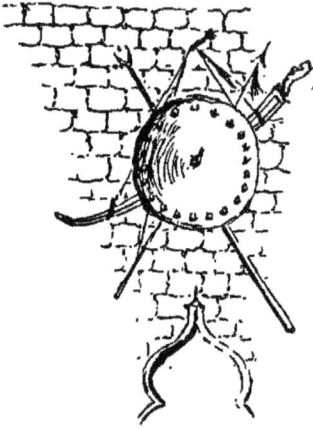

THE Political character of the Pyrenees corresponds to the Physical character which has been described. The high crest is the bond and division, from the beginning, between two societies which are connected by such common social habits as mountains impose—which therefore fall under similar local customs, which have a common jealousy of the civilised power on the plains below them, and which support each other in a tacit way against the stranger, yet which, from the beginning, have different governments and (especially in the high central part) deal with different corporate traditions—to the north the Bearnese, to the south Aragon. The easier passes to the west and the east of the chain permit a more or less homogeneous community to straddle across either end of the mountains, and to hold upon both slopes the sea roads that pass along the Atlantic and the Mediterranean. The people thus astraddle of the eastern end have come to be called the *Catalans*. That astraddle of the western, a highly distinct group of men with language, traditions and physical characteristics wholly their own, has always been known by some title closely resembling their modern name of *Basques*.

The foundation, therefore, upon which Pyrenean History is built, or (to use another metaphor), the germ from which it has developed and which explains its course, is a tripartite division of the inhabitants, corresponding, as I shall presently show, to the physical features of the chain : an eastern or Catalan, a western or Basque, and a central group whose characteristic it is to sub-divide according to the deep valleys into which it is separated, but which falls into two main societies, the one north of the chain which becomes the group of French counties whose typical government is Bearn, the other south of the chain, which assumes at last for its title " The Kingdom of Aragon."

The first matter to be noticed with regard to this tripartite division is the exactitude of its boundaries. One might imagine that the language, the habits, and the clear characteristics of any group would merge easily into those of its neighbours upon either side. This is not the case. The Basque type—much the most particular—ceases abruptly upon the watershed between the Gave d'Oloron and the Gave d'Aspe to the north of the range, upon the watershed between the Veral and the Esca to the south of it. The Catalans, with a dialect, mind, and dress wholly their own, are found to the *north* from the sea up to the Col de Puymorens, and everywhere east of the Carlitte mountains ; in the Ariège valley and just over these heights, and on the further side of that Col, they are changed. To the *south* of the range they extend everywhere from the sea to the valley of the Ribargorza. Cross westward from that Catalan valley to the Esera. There, after hours of scrambling, down by the rocks and deserted tarns, you may towards evening find a man ; that man will show the slow gestures, the silence, and the elaborate courtesy of Aragon.

The mountain ridges which divide these various peoples are sufficient to mark their boundaries ; but they do not suffice to explain why the Catalan, the Basque, the Aragonese, the Bearnais should cease suddenly here or there. True, the high lateral ridges which are so striking a peculiarity of the Pyrenees form barriers with difficulty passed, but these barriers are found just as high and just as precipitous and savage between two valleys of the same speech and nation as between two of different allegiance. Thus the wild jumble of mountains, "the Enchanted Range," cuts off the Catalans of Esterri from the Catalans of the Ribargorzana. To pass them is something of a feat for anyone not of these hills —for much of the year they are closed to the native inhabitants. Their passage is hardly more of a task or more precipitous than the passage from Aragonese Venasque to Aragonese Bielsa, or from Bearnais Gabas, in the Val d'Ossau, to Bearnais Urdos, in the Val d'Aspe.

An explanation of the unity which rules over each group, Basque, Central and Catalan, can only

The Pic d'Anie from the Bridge of Oloron

THE PIC D'ANIE FROM OLORON

be given by referring each to the plains at the mouths of the valleys. It is the towns at the entry of these plains that form the markets and rallying places of the mountaineers and that determine their groupings. Oloron is the link between the two Bearnais valleys I have mentioned. Urgel binds Catalan Andorra to Catalan Esterri. Why, however, the groups should lie exactly where they do it is impossible to determine, for no records reach beyond the Romans. All we can say is that the Pic d'Anie, the first high peak eastward from the Mediterranean, forms the boundary stone of the Basques, as it does the chief physical mark dividing the high central ridge from the easier western passes; that the tangle of difficult and impossible peaks just eastward of the Maladetta are the boundary of Catalan south of the range, the similar but less abrupt tangle of the Carlitte, their boundary upon the north. How these nations arose, whence they wandered, whether their differentiation has arisen upon the spot out of an earlier homogeneity or is due to the conflict of invaders—of all this we know nothing.

The place names of the Pyrenees, like those of all Spain, and half Gascony, do indeed afford a curious speculation which arises from the high proportion of names that are certainly *Basque*, though out of Basque territory. Of this language I shall write later: for my present purpose the point I would desire the reader to note is the sharp contrast which exists between that idiom and the idioms around it. There is no mistaking a Basque word, and yet these are found in all the Pyrenean range and to the north and south of them in a hundred place names, attached to hills, rivers and towns where Basque has been unknown throughout all recorded history. It is even plausibly suggested that the Latin "Vascones," the French "Gascon" is equivalent to "Basque," and the late Mr York Powell, the Regius Professor of History at Oxford, would say in speaking upon this matter that "Gascon was Latin spoken by Basques." He possessed that type of education, rare or unknown in our universities, which made him capable of individual judgment in departments of living knowledge where his colleagues could but repeat words taught them from a book. This quality reposed upon a wide acquaintance with all matters of European interest His diverse reading and considerable travel enabled him to balance human evidence in a way hopeless to his less fortunate neighbours in the University, and his conclusion on this important detail of history has always recurred to me when I have examined some new point in the early history of these mountains. There must, however, be set against the general conclusion that the Basques are the remnant of a people once universal from the Garonne to the Pyrenees, and throughout the Iberian Peninsula, the fact that they present a marked physical type utterly distinct from others upon every side. That a race of such a character, vigorous, attached to the soil, in no way nomadic, should have abandoned a large territory is difficult to believe; moreover, there is no case in all the recorded history of Western Europe of one people ousting another, and the process is manifestly physically impossible, save among nomads. Jews or Arabs could propagate and even believe such a theory. To Europeans it is laughable : the peasants and cities of Europe never have been, nor ever can be, largely displaced.

All we know is that these place names exist throughout Spain and all over the Pyrenees, and that the million or so who speak the language whence such names are derived now occupy a tiny corner only of the vast territory over which those names are spread. The rest is guesswork.

Ignorant as we are of the origin of the differentiation between Basque, Bearnais, Catalan and Aragonese, an historical fact quite certain—though no document proves it—is the extreme antiquity of these classes of men. That all Pyrenean history reposes upon their separate existence must be evident to anyone who has watched the commercial manner, the mercantile vivacity, the whole mentality of the Catalan, and has contrasted it with the quiet chivalry of Aragon. Different military fortunes, different economic outlets, and different accidents of central government may possibly account within the historic period for the contrast between the Aragonese and the people of Bearn, Bigorre or Comminges. No such forces can account for the gulf that cuts off the Catalan and the Basque at either end of the chain from the inhabitants of its high central portion. Infinite time is the maker of states, and two thousand years could never have determined societies so sharply separate. We must regard their constant and immemorial presence in the Pyrenees as the first and enduring principle to guide us in the history of those mountains.

From this fundamental truth, which leads the prehistoric into the historic, one must proceed to another political fact of high importance, which is that while the watershed of the range has but partially separated customs and local thought, and that only in the centre of the range, it has necessarily served as a political boundary whenever a high civilisation found it necessary to establish such a strict line. The boundary and the watershed may not exactly coincide—they do not exactly coincide even in the highly organised condition of modern society ; but in the two historical periods of strict policy, the Roman and our own, the crest of the range has marked, and marks, an obvious boundary for most of its length. The political distinction between Hispania and Gaul cut the Basque nation into two, following the mountains from Roncesvalles to the Pic d'Anie : it cut the Catalan people into two, following the water parting from the two Nogueras to the Mediterranean. It followed the central chain, indifferent to the similarity or difference between the northern and the southern valleys. To-day the political distinction between Spain and France follows nearly the same line.

The reason of this was, and is, twofold. First, that a clear physical boundary easily definable and of its nature permanent—the crest of a chain, a broad river, or what not—necessarily recommends itself to a bureaucracy in search of simplicity and economy in the work of a great political machine. We see it in the new countries to-day, where the instinct of organised government for easily definable and exact limits takes refuge in establishing parallels of latitude as state boundaries in the absence of marked physical lines. Secondly, in the case of mountains, and especially of mountains as sharp and as boldly set as are the Pyrenees, the fatigue of climbing, the absence of carriageways, made each valley dependent for its connection with the central government upon some town of the plains, and the authority of a provincial magistrate could not but run, as ran the physical instruments of his rule, up from Huesca northward to Sallent—for instance, or up from Jaca to Canfranc, and so to the summit of the ridge; or up from Oloron southward to Accous, and so to Urdos. As the messengers, writs, powers of each proceeded, the way would become harder, the progress more doubtful. It was obvious and necessary that the boundary of either jurisdiction should lie upon the pass. And though the inhabitants of the northern and the southern valleys might be accustomed to a regular intercourse across the crest, the Roman agents of a distant central government could not but have depended upon cities far removed to the south and to the north of the watershed, as to-day the police of Tardets, let us say, and of Isaba, two towns of one speech, refer respectively through Pau and through Pamplona to Paris and to Madrid.

It is in the interplay of these two jarring political

forces, the permanent national seats of Basque, Catalan, etc., and the use of the range as a political or official boundary, that the political character of the Pyrenees resides ; and as their history begins with the Romans, to whom we owe the first knowledge of the Pyrenean people and the first use of the Pyrenean boundary, it will be well to consider it under territories divided as the Romans divided them, by the main range, and to follow first the development of the northern slope.

The historical origins of the French Pyrenees are sharply divided in history by that wall which cuts off all that Rome *made* from all that Rome *inherited.* Rome made of the barbarians a new world, but before she began that task Rome had inherited everywhere within a march of the Mediterranean a belt of land whose civilisation was similar to, always as old as, and sometimes older than, her own. It was a municipal civilisation dependent upon the arts and religion proper to a city state. It built, whether temples or ships, as Rome would build them : it was one thing ; it is almost one thing to-day ; and its bond at Antioch as at Saguntum, at Marseilles as at Athens or Alexandria, was, and is, the universal water of the Mediterranean To such cities and their territories Rome fell heir. Little proceeded from her to them save first the sense of unity, and later the Faith, and of the whole system, the belt which stretches from Valencia to Genoa, now broadening to the plains of Nemosus (Nîmes), now narrowing to the rocky ledge of the Portus Veneris (Port Vendres), concerns the first evidence of Pyrenean history; for it was from a corner of this belt—between Tarragona and Narbonne—that the advance of civilisation inland and along the Chain proceeded.

A century before the four imperial centuries which made our Christian world, a century before Augustus Cæsar, Rome had fully occupied and impressed that soil—to the south Gerona and the Catalan fields, to the north the rich floor which lies under the Canigou and has come to be called the *Rousillon*. Thence the Roman advance north of the hills proceeded. The chief town of the sea-plain—whose name "Illiberis" is so strongly Basque in form—Rome took for the central municipality of that plain, and made it the capital of the coastal district. This hill and citadel, at which Hannibal had halted a hundred years before, preserved as a bishopric for thirteen hundred years a memory of the Roman order. Constantine formed its diocese, rebuilt it, gave it his mother's name of Helena. The sea by which it lived has withdrawn from it. It has sunk to be a little country town, "Elne." Roscino, which lay also upon the coast march of Hannibal, has sunk to something smaller still, yet, by some accident, gave the province, in the dark ages, its name of Rousillon which it still retains. These two towns, the fruitful plain about them, the Port of Venus (which is now Port Vendres), formed the municipal structure of this district, the last corner of the great province whose headship lay at Narbonne. Its nominal boundaries included all the vale of the

Tet ; it extended as far as now extends the Catalan language, and was bounded, as that is bounded, by the great form of the Carlitte and its high lakes and snows. All between that mountain and the sea, all the eastern decline of the range and the slope north of it, was ancient land, and had been ploughed and held and walled by men of the Mediterranean civilisation long before Rome inherited it.

With the much longer stretch that runs from the upper Ariège to the Atlantic it was very different. This was of what Rome made, not of what Rome inherited. Before the coming of Roman government it was barbarous, and the many tribes or petty states, whose number various guesses of antiquity record (they were perhaps as numerous in their subdivision as the valleys), stand in three main groups when first the civilisation to the east of them began to record their existence : these three were first the Convenæ, south of Toulouse, and all about the upper waters of the Garonne. Next to these came the Auscians, and finally, over the Basque end of thè hills towards the ocean, was the seat of the Tarbelli.

The whole point of view of antiquity differed from ours in speaking of such tribes, nor is it easy to pick out from the scraps of observation that have come down to us the kind of information that we want. Sometimes a name survives, sometimes it does not ; sometimes we get a hint of a variety of race, most often we lack it. It is the very meagreness and eccentricity of the information upon a barbarous race and custom which affords such opportunities to our dons for those forms of speculation which they love to put forward as dogma, the most absurd example of which, perhaps, is the interpretation and enlargement of Tacitus' "Germania." It is therefore exceedingly difficult to know of what kind were these people beyond the old Roman pale. We do not know what language they spoke. We only know that, like other Gallic communities, they centred round fortified places, that their pacification was easy, and that, like everything else in Western Europe, they were of an unchangeable kind.

The whole district between the Garonne and the Pyrenees came to be called, during the first four centuries of our era, " The Nine Peoples." The Convenæ are early noted to have attached to them upon their right and upon their left, to east and to west, the Consovanni and the Bigerriones. The first of these were (to follow the high authority of Duchesne) organised as early as the first century; what is now St Lizier was their old capital and later their bishopric, which takes its present name from Glycerius, a Saint of the sixth century. They held all those hills of which St Girons close by is now the centre. The Bigerriones are not heard of until the mention of them in the Notitia of the fourth century. They must have held Bigorre, and the three valleys which I have called the valleys of Tarbes. Tarbes—then Turba—was their capital, and was and is their bishopric.

The Auscians do not concern us. They and the three groups into which they are later distinguished held the western plains and foot hills. The Tarbelli held both the foot hills and the mountains of the west; their capital was at Dax. They also split into, or are later recognised as three separate groups, making up with the two other sets of three "The Nine Peoples," under which title all this country below the Pyrenees became permanently known. But of the three only the *Civitas Benarnensium*, whence we get the name Bèarn, and the *Civitas Elloronensium*, with its capital at Iloro, which has become Oloron, concern us. The capital and soon the bishopric of the *Civitas Benarnensium* was at Lescar, as far as we can make out, and Lescar bore the chief sanctity in Bèarn until that country was swept by the Reformation. The sovereigns of Bèarn were buried there, even the Protestant sovereigns, and it remained a bishopric, whose bishop was the President of the Parliament of Béarn, until the Revolution ; but it was the Reformation which destroyed its original character of a capital.

We have therefore, with the earliest ages of our civilisation, five peoples holding the northern Pyrenees, the Consevanni, the Convenæ, the people of Bigorre, the Bèarnese, and the Elloronians.

It is remarkable that in such a list, our Roman originators and their geographers overlooked the Basques. The category ends precisely at the present limit of the Basque tongue. For the Val d'Aspe, of which Oloron is the town, is the first French-speaking valley. Why it is that we hear nothing of the Basques it is difficult to say, especially as the second of the great Roman military roads went right through their country. Bayonne, which is the Basque's town of the plains on the north, is heard of in the fifth century. It has a garrison; but no bishopric until the tenth. Pamplona, which is their town on the south, was known before the beginnings of our Christian history. But the Basques themselves are not known to us from the Romans. The name of the Consovanni survives locally. The country round St Girons is still all one countryside and called the "Conserans." Of the Convenæ we have a pleasant legend in St Jerome telling how Pompey got together all the brigands of the mountains, drove them northward hither, and forced them into a garrison (a stronghold which, like Lyons and the rest, was one of the many "Lugdunums"). It was destroyed early in the dark ages, and later revived by St Bertrand, a little way off in his Episcopal town. Their name survives in the district of Comminges. The Bèarnese name of course survives and so does the Bigorrean, while the Elloronean, though no longer the title of a district, is preserved in the town name of Oloron.

All this country, not only that of the five tribes along the mountains, but the whole territory occupied by the nine peoples (who afterwards became twelve), lay in a profound peace under Roman rule, and we may be certain of its increasing wealth throughout the first four great formative centuries of our era.

The advance of Rome upon the Spanish side was of a very different kind. Rome, after the Carthaginian wars, inherited broad belts of civilised and half civilised land. All the Mediterranean slope below the mouth of the Ebro, and a belt quite three days' marches wide inland to the north of that river, was full of ancient populated towns, alive with the full civilisation common to every shore of the inland sea. So, we may be certain, were the broad plains of the south where the most complete and earliest absorption of the Celtiberian in Roman speech and ideas took place. The advance into the north-west and therefore along the Pyrenees covered more than a century of strict and perpetual warfare, which was intermixed by the civil wars of the Roman commanders. The extremities of the Asturias were reached in the century before the birth of our Lord, but the advance was not, as upon the north, a rapid expansion beyond the old boundaries. It took the form of siege after siege and battle after battle, in which those numerous and crushing defeats, which Rome (like every truly military power) reckoned to be a necessary part of history, interrupted the slow progress of her law. The Celt-Iberian towns were walled and strong; their resistance was painful and tenacious; there was no sudden illumination of a willing people by a new culture, such as had taken place in Gaul, rather was northern Spain kneaded by generations of warfare into the stuff of the Empire.

When the work was accomplished, it was complete throughout the Peninsula; and though the silent strength of the Basques prevented the Roman language from invading their valleys, the administration of the whole territory south of the

Pyrenees must have been as exact and as bureaucratic as that to the north of it. There was, however, this great difference due to local topography between the Spanish and the French hills, that the municipalities upon which Rome stretched her power, as upon pegs, were less common, were farther apart, and approached less nearly to the central ridge upon the southern than upon the northern side. What you see to-day south of the Pyrenees is what you might have seen at any time in the last 2000 years—a very few scattered towns, still the centres of government, and all the rest rare isolated villages living their own life, free from the criminal, and, by a regular payment of small taxes, half independent of the civil, law. Alone of the true mountain sites, Jaca in the middle, Pamplona and Urgel at either extremity, were bishoprics. Huesca, St Laurence's town, a fourth centre, is in the plains. For the rest the confused storm of hills ending in those parallel ranges, pushed right out on to the burnt flats of the Ebro, forbade the establishment of a municipal civilisation.

Upon all this land to the north and to the south of the mountains came, after five hundred years of a high civilisation, the slow decline of culture, and the infiltration of the barbarians. In a sense the nominal divisions between the barbaric kingdoms has its importance, for they help us to understand changes of dynasty and of custom. But they were of no political effect. The mass of the people knew little of the chance soldiers who, with their mixed retinues of Roman, Breton, German, Slav, and the rest—some able, some not able to read the letters of Rome—sat in the old seats of office, issued their writs through the still surviving Roman Bureaucracy and from palaces which were but those of decayed Roman governors.

For the greater part of Western Europe, and especially of Gaul, this process of decay was one into which Europe slowly dipped as into a bath of sleep, and out of which it rose more rapidly through the energy of the Crusades and of the renewed Pontificate into the splendour of the Middle Ages. But the Pyrenees suffered in this matter a peculiar fate. When Spain was overrun by the Mohammedan, and when in the first generation of the eighth century the Asiatic with his alien creed and morals had even swept for a moment into Gaul, the Pyrenees became a march : at first the rampart, later, when they were fully held, the bastion of our civilisation against its chief peril. It is this episode by which the Pyrenees became the military base of the advance against Islam—an episode covering the whole life of Charlemagne and after him the ninth, tenth, and eleventh centuries—which gives them their legendary atmosphere and fills all their names with romance.

The northern slope, during the long business by which Gaul became itself again, was but a remote border province. The new life of Gaul, after the shock which had so nearly destroyed Europe was over, sprang from Paris. The influence of Paris radiating upon every side built up again accuracy of knowledge, unity of government, and general law. To this influence the Pyrenees seemed the most remote of boundaries. The valleys were little affected by the growth of the French Monarchy; they remained for centuries broken into a maze of half-republican customs, of tiny independent lordships, guarded and menaced by separate and jealous walled municipalities upon the plains—all of this vaguely and slowly coalesced into larger districts, doubtful of their sovereignty and perpetually struggling upon their boundaries and their sub-boundaries.

In this development nothing was more striking than the way in which this remote border at first looked rather to the south, where the interest of religious war was ever present, than to the north, whence government was slowly coming towards it. The French Pyrenees fought and felt with the whole range, not with the plains. Jaca in the worst time, when it was the only mountain bishopric free from the Mahommedan, counselled with and was perhaps suffragan to Eauze. Urgel sat in the provincial Synod of Narbonne. As the success of the Reconquista pushed the noise of crusade further and further from the range, the northern valleys looked more and more towards their northern towns. Their nominal allegiances grew stricter—as of Foix to Toulouse—and every French bishopric was bound more and more to its northern metropolitan, the Spanish sees to the new metropolitans of the Ebro.

At last an issue was joined between Northern and Southern France of the first moment to the unity of Gaul itself and of all Christendom. The

Crusades, the knowledge of the East, the awakening of the intelligence and of its appetites, had bred throughout the wealthy towns of what had been from the beginning Roman land, a desire to be rid of the restraints of fixed religion. The South of France began to move towards its pagan past. It was a movement which had already had its strange echo in the north, a movement which in England had only been pulled up at the last moment by the martyrdom of St Thomas. Here in Gaul, in the sunlight, and backed by so much gold, the rational and sensual revolt became a larger thing, and when various sources of disruption, speculation, and achievement had met in one stream, it was commonly called the *Albigensian* movement. The issue was decided, after heavy fighting, in the early thirteenth century, and the victory was with the cause of the unity of Europe. Toulouse (the true centre of the storm) and its lord were conquered. Northern barons swept down, held no small part of the southern land, and from that time onward the French Pyrenees are normally dependent on Paris.

Two exceptions survived, the straddle of the Basque across the chain and the straddle of the Catalan. The Basque had his country of Navarre upon either side of the chain; with it went Bèarn, and these were independent of the French crown. The Catalan, able to traverse the chain by the flat floor of the Cerdagne, preserved the unity of his mountain province, and the Roussillon counted with Spain. Apart from this easy passage into the Roussillon from the south, by way of the

Cerdagne, the isolation of the Roussillon was the more easily accomplished from the long spur of the Corbieres, which runs north and east towards the sea and cuts off from France the wealthy plain of which Elne had once been, Perpignan had later become, the capital.

This arrangement endured, in name at least, until the seventeenth century. The last heir of Navarre was also the heir nearest to the French throne at the close of the religious wars, and as Henry IV. of France he united the two crowns. A man who, as a boy, might have rejoiced at that union, could have lived to see, under Mazarin, the signing of the Treaty of the Pyrenees, which gave the Roussillon also to the French Crown. The date of this final arrangement coincided with what is ironically called "The Restoration" in England · this date, which definitely closed the power of the English Monarchy, and substituted for it the power of a wealthy oligarchic class, coincides throughout Europe with the struggle between absolute central government for the equal service of all, and local aristocratic custom. In England the latter conquered ; in Spain and France the debate was decided in favour of the former.

Such centralised governments could but further define and insist upon a new boundary, and from that time onward, for 250 years that is, the Pyrenees have been once more as they were under the clear administration of Rome, a fixed political boundary ; and, save certain exceptions that will be mentioned later, everything north of the chain has been ad-

ministered from Paris, and has slowly accepted, side
by side with the local tongue, the tongue of Northern
France and the habit of a centralised. French
government.

South of the chain the process by which Christendom recrystallised out of the flux of the dark ages,
followed a different course; it was a process to
which Spain owes all her national characteristics,
for out of the mountains a Spanish nation was
formed, and from its various communities, as from
roots, the Catholic kingships grew southward until
they once more reached Gibraltar.

To understand this process, it is necessary to
consider factors absent in the topography of the
Gaulish plains, and especially the factor of that un-
conquerable tangle of mountains which occupies all
the north-western triangle of the Peninsula.

The ocean boundaries of the Iberian quadrilateral
are nearly square to the points of the compass. It
is not so with the internal divisions that mark off its
central part. Here the edges of the high and arid
plateaux, the deep trenches of the rivers, the
mountain ranges, the boundaries of the plains at
their feet, run slantways from north-east to the south-
west. This slant determined the boundaries of
Mohammedan expansion, while the Asiatics and
Africans still retained the energy to advance; it
determined the successive frontiers of the Re-
Conquest, as our race slowly ousted the invader and
reached at last the sea-coast of Granada. The
Arab and the Moor were masters of Narbonne and
all the Roussillon on the east, when, on the west, they

could not cross the mouth of the Mulio a hundred miles to the south. They were at Jaca within a day's march of the watershed along the Roman Road, when, to the immediate west of it, they could not hold Fuente ; they could not even reach Pamplona, though that western town is two marches at least from the main crest. Toledo was reconquered a generation before Saragossa, though Saragossa is by

nearly two degrees more northerly, because Toledo was west of Saragossa. The last Mohammedan kingdom was crowded, after the thirteenth century, into the extreme south-east, as the surviving remnant of the free Europeans of the Peninsula had been crowded into the extreme north-west in the eighth.

If the boundaries of undisputed Mohammedan rule be traced for various dates, the receding wave will be found in general to follow curves that lead,

like the main features of the land, from the north-
east downwards towards the Atlantic.

This main character in the geography and history
of Spain, the south-westerly trend of the mountain
ridges, largely determined the fortunes of those
fighting bands of mountaineers who ceaselessly
pressed southward until they had wholly driven out
the invader and reconstituted the unity of Europe.
It determined the first advance to be, not from the
Pyrenees, but from the Asturias, and the first captain
connected with the Christian resistance after the
overwhelming of all that civilisation, *Pelayo* (from
whose blood Leon, Castille, Aragon, and Navarre
descend) had his stronghold, not in the Pyrenees,
but a week's march to the west, along the Biscayan
coast at Canzas. Within the decade of the invasion
he had checked the invader in his own hills at
Cavadonga.

All the eighth century is full of that successful
spirit in the north-west—but nowhere else. Alfonso,
the husband of Pelayo's daughter, struck the note
with his boast, " No pact with the infidel," and the
tradition or prophecy that Christendom would regain
the south, springs from him. He conquered down
to the Douro, over what is to-day the mountain
frontier of Portugal ; he began those long cavalry
raids into the heart of Moorish land. He rode into
Astorga, into Zamora, into Segovia itself—within
sight of the central range of the Guadarama : riding
back with booty, harassed and harassing, nowhere
permanently fixing himself save in the towns of the
west, upon the Lower Douro, but building on the

ridges of his defence, those block-houses, the "Castille" from which, long after, the frontier province began to take its name.

All the ninth century that spirit grew. The body of St James was found under the Star at Compostella —its shrine became the national sacrament as it were, a perpetual refreshment for arms, and a symbol, in its wild isolation among the rocks of Gallicia, of the impregnable places from which the Reconquista drew its ardour. The advance continued. The frontier counties consolidated and were named.

Leon was permanently held, Burgos was founded. If one takes for a date the opening years of the tenth century, just after Alfred had saved England also from the pagan, and just after the Counts of Paris had saved northern Gaul, there is a full Spanish kingdom standing up against the Mohammedan power, a king has been crowned in Leon and has died in peace at Zamora. The cavalry raids have pushed—once at least— to Toledo. All the north-west lay permanently Christian beyond a line that ran from the corner of Gaul to the Douro and down the Douro to the sea ; and this united triangle of Roman land formed a base from which the pushing back of the alien could proceed.

How did this disposition of forces affect the Pyrenees? Let it first be noted that the newly organised Christian country lay wholly to the west of the range. In the Pyrenees themselves the Mohammedan flood had washed every valley.

The crest had been traversed and retraversed ; both slopes were for a moment held by the invaders. Abd-ur-rahman had sent or led his thousands by the Roman roads of Roncesvalles and Urdos and over the Ostondo and the lower passes of the west. The mule tracks of these rocks had been twice crowded with the white cloaks of the Arabs. In the east, Narbonne was held for fifty years, and with it all the Catalans. Even in the high centre of the chain, where there is no passing between the Somport and the Cerdagne, wherever there was something to rob or to destroy, the invaders had penetrated. There was not here, as in the Asturias, untouched land.

When the crest of the wave retreated, when the Mohammedan came back defeated from Gaul, the high valleys attained—it may be guessed—a savage independence.

Jaca has legends of its battle at the very beginning of the Independence, before Charlemagne had come to the rescue, and from all the valleys of the Sobrarbe, bands of men must have been perpetually volunteering for skirmishes down into the plains. Navarre was the natural leader of the movement, the largest and the most fertile belt of Christian land, but the little lordship upon the Aragon, fighting down south and east towards the Ebro, the western count of the Asturias fighting down south and west, cut off the advance of the Basques ; and though Navarre in the period of birth and turmoil which is that of Gregory VII.'s reform of the Papacy, of the establishment in England and in

Sicily of Norman power, and of preparation for the Crusade, was the head of all the southern Pyrenees and called itself an " Empire," it was blocked by the double line of advance, and the Basques, upon the foundation of whose tenacity and courage, as upon a pivot, the Reconquest had proceeded, took little more part in the wars ; but the Basque stirp of Navarre gave its first king to Aragon, and the son of that first king, Sancho, raided so far as Huesca and was killed beneath its walls ; his son again, Peter, took the town two years later just as the hosts of Europe were gathering for that first great march upon Jerusalem which threw open the curtain of the Middle Ages ; and *his* son, Alfonso (who had united in one crown Leon and Aragon), went forward under his great name of the Batallador, and twenty years later (1119) swept into Saragossa, the last of the Mohammedan strongholds in the north.

Thus were the west and the centre of the Spanish slope recovered for our race and civilisation.

Meanwhile Catalonia upon the east had been since Charlemagne, since the early ninth century, a march of Christendom ; but it was not until the same creative period which had brought forth the leadership of Navarre and the advance from Aragon, the Normans, Hildebrand, and the resurrection of Europe, that Catalonia began to go forward. Its first true monarch was Berenguer the Old, who lived round and about the date of Hastings, and was first master of the whole province. He also founded and maintained the Cortes of Barcelona. His son, for a moment, raided the Balearics, and when he died Catalonia and Aragon, united under one crown, saw the alien finally driven from these mountains. All the plain from far beyond the base of the hills was now permanently held by strong and united kingships which pressed forward to the Ebro valley, and finally saved all the Spanish province of Europe. A lifetime later, the last of the foreign armies had been broken at Navas de Tolosa. Far off in the south Islam lingered, tolerated and on sufferance, but Spain was reconquered. For just 500 years Spain, a quarter of all that makes up our civilisation, had lain in peril between our religion and the other.

I have said that with the thirteenth century, the Albigensian crusade upon the north, the destruction of Islam upon the south (the two successes were contemporaneous), the Pyrenees ceased forever to be a march between two civilisations, and became a mere political boundary between two provinces of Europe ; and I have said that the nature of that boundary was finally fixed in the seventeenth century, or rather during a period which stretches from the close of the sixteenth to just after the middle of the seventeenth.

If that political boundary be examined to-day it will be found to coincide with the watershed, save at certain particular points, the character of which merits examination.

I take for my boundaries, as throughout this book, Mount Urtioga on the west, and the beginning of

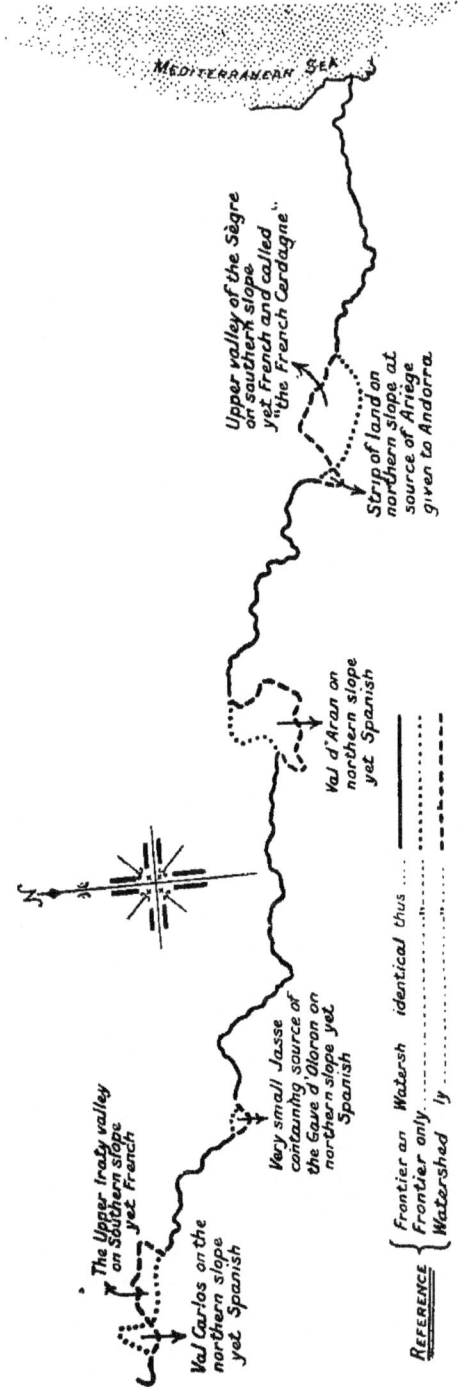

MEDITERRANEAN SEA

Upper valley of the Sègre on southern slope yet French and called "the French Cerdagne"

Strip of land on northern slope at source of Ariège given to Andorra

Val d'Aran on northern slope yet Spanish

The Upper Treaty valley on Southern slope yet French

Val Carlos on the northern slope yet Spanish

Very small Jasse containing source of the Gave d'Oloron on northern slope yet Spanish

REFERENCE {
Frontier an Watersh identical thus ....
Frontier only ....
Watershed ly ....
}

PLAN K

the Alberes on the east, which may conveniently be placed at the Couloum.

In this distance, there is a slight discrepancy between the political boundary and the watershed here and there in the Basque valleys. Mount Urtioga itself, though upon the watershed, is entirely in Spain, and the sources of the torrents which feed the valley of Baigorry all rise a mile or so beyond the political frontier, which is here composed of two straight conventional lines.

The head waters of the Nive are wholly in Spain also, as is all the left bank of that river to a point four miles below Val Carlos. The right bank, however, is French, so far as the torrent Garratono. Thence forward from the sources of that torrent (that is, upon the Atheta) the frontier now follows the watershed, now leaves the very head-springs of the torrents in Spain until, a few miles further east, it makes a considerable invasion of Spanish territory, not because the frontier itself bends, but because the watershed here goes northward in a half circle. All the upper valley of the Iraty is politically in France ; but from the Pic d'Orhy, where a definite ridge begins, it follows the frontier strictly for mile after mile (with the exception of a curious little enclave which gives Spain two or three hundred yards of the head-waters of the Aspe), and there is no further exception throughout all the high Pyrenees until one strikes the curious anomaly of the *Val d' Aran.*

I have said in describing the physical structure of the Pyrenees that the two main axes of those mountains were joined by a sort of fault, a serpentine bridge of high land which united them from the Sabouredo to a point ten miles northward, the Pic De l'Homme overhanging the Pass of Bonaigo. The valley caught on the French side of this twist is the Val d'Aran, containing the upper waters of the French river Garonne.   Geographically of course it is French, but politically it is Spanish so far as a certain gorge where is a bridge called the King's Bridge, and where the Garonne pours through a narrow gate of rock into its lower valley.   The story goes that when the Treaty of the Pyrenees was in act of negotiation some one said diplomatically and casually to the French negotiators, " The Val d'Aran of course you regard as Spanish," and they, knowing no more of these mountains than of the mountains of the moon, said, " Of course."   The true reason is rather that the gate in the mountains cuts off this upper valley from the lower gorges of the river much more than the low, easy, and grassy saddle of the Boniaigo cuts it off from the Spanish valley of Eneou just to the east of it ; and though the Val d'Aran may be geographically or rather hydrographically French, it is topographically Spanish, which is as though one were to say that Almighty God made it so.

Another exception and a big one to the rule that the frontier follows the watershed, is, of course, to be found in the French Cerdagne.   The true watershed here is coincident with the frontier as far as the Pic de la Cabanette in latitude $42° 35' 30''$.   The watershed then goes on over the Port de Saldeu,

along the crest of the Port d'Embalire to the Pic
Nègre, and there it turns to the east along the ridge
across the saddle of which goes the high road over
the Col called Puymorens. It follows that ridge,
not to the summit of the Carlitte, but to a lower
peak called the Madides, three miles to the north-
east, runs along two miles of a high rocky ledge to
the Pic de la Madge. and then there follows a difficult
sort of hydrographical No Man's Land, the centre
of which is the great marsh of Pouillouse, nor
can you tell exactly where the watershed is for
some miles in the forest below that marsh, for
the same damp flat ground sends water into the
valley of the Tet and into the valley of the Segre.
Three miles to the south-west, however, it is clearly
defined again in a low rounded lump of wooded
land, it passes over the flat Col de la Perche and then
follows the crest still going south-west up to the Pic
d'Eyne, where again it becomes the frontier, and the
frontier it remains until it reaches the Mediterranean.

From the Pic de la Cabanette, all the way to the
Pic d'Eyne, France and Mazarin politically took in
by the Treaty of the Pyrenees a belt to the south
of the watershed and extending down to a conventional line which left Bourg-Madame French
and Puigcerda Spanish; an exception in this is
a small strip beyond the Pic de la Cabanette,
on the left bank of the Ariège, which, though
geographically French, was given to Andorra,
so that Andorra might smuggle more comfortably
over the passes.

The causes of this annexation of the French

THE CULMINATING POINT OF THE RANGE

Cerdagne by Mazarin are clear enough when one remembers that the Roussillon (which is geographically French) passed to France by the same treaty. There is no way from the valley of the Ariège into the Roussillon except by going round this corner of the Cerdagne, at least no practicable carriage way; the only other way is the difficult and high short cut described later in this book. If the frontier be carefully noted, it will be seen that it is designed merely to preserve a right of passage over this road. Jurisdiction was only claimed by France over the villages, and Llivia, being a town, stands in an island of Spanish territory in the midst of the French Cerdagne, as will be seen later when I speak of this district in detail.

Such is the present political aspect of the Pyrenees, with Toulouse for their great French town in the plains, 60 miles away to the north, Saragossa for their great Spanish town in the plains, 100 miles away to the south, a string of towns just at their feet (Bayonne, Pau, Tarbes, St Girons, etc.) on the northern side; on the south a rarer and less connected group, (Pamplona, Huesca, Barbastro, Lerida, etc.); and against the Mediterranean the district of Gerona, shut in by the Sierra del Cadi (with its outposts) and the Alberes upon the Spanish side, the town of Gerona its capital; the Roussillon, with Perpignan for its capital, shut in between the Alberes and the Corbieres on the French side.

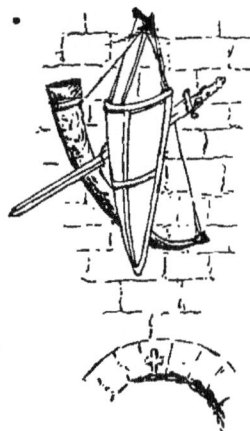

III

MAPS

ONE of the first ideas that come to a man when he thinks of wandering about an unknown bit of country is that it will be more fun if he does not take a map. There are places of which this is true : you discover for yourself, and it is more exciting. But it is not true of the Pyrenees. So little is it true of the Pyrenees that those who have no maps, that is, the local peasantry, never traverse a country until they know it well, and when they get into new country learn all they can from its inhabitants, get themselves accompanied if possible, and keep to a path. You will find that the hunters who know the mountains are always local men. The Pyrenees are built in such a fashion and on such a scale that you not only can, but must lose yourself in the course of any long wandering unless you have some sort of guide to your hand. There is only one kind of travel off the road which you can possibly undertake without a map, and that will be pottering about one small district with a porter, a friend, or a mule to carry a tent and plenty of provisions ; but if you are attempting several crossings of the ridges, and especially if you are attempting such a task on foot, a map is absolutely necessary to you.

Whatever kind of map you take with you into the hills, you must also take with you a small compass, and that is why I mention that toy later in talking of equipment. You are perpetually asking yourself, as you compare the map with the landscape, which peak is which, and it is often essential to get the right one on the right bearings. Nothing is easier than to mistake one part of a ridge for another.

If you are in bad weather or in the dark or enclosed, the compass gives you a general direction, as for instance upon the track I describe later in the great wood going to Foumigières, and the compass further tells you at what point your valley begins to turn in a certain direction. Now a bend of this sort is very often the only indication you have for the exact place in which to branch off for a port, or to look for a cabane. Remember the variation, which is on the average for this range about 14 degrees, that is, the true north is 14 degrees to the right of the direction the needle points to.

A map or maps, then, you must determine to take, and it next remains to examine what sort of maps are available for the whole range.

There are but three of the greater countries in the whole world (to my knowledge, at least) which have sufficient and numerous maps, these are England, France, and Germany. I can imagine what reproach and criticism such a statement may bring from those who know the admirable work done in India, and the special but laborious surveys of Italy and of the United States. But I do say

6

(as far as my travels extend) that maps valuable for the purposes of a man on foot and covering a whole country are confined to these three among the greater states. To tell the truth, there is but one large country that possesses perfect ones, and that is our own. Nowhere else in the world (to my knowledge, at least) has a complete survey of every detail of the soil been made, as it has been made, under the Crown of the United Kingdom. And if foreigners judge, as they are apt to judge, of our cartography by the excellent one-inch scale map alone, they should remember that we also possess the six-inch, and in some cases the twenty-five inch to supplement it. Neither France nor Germany can boast of such a survey.

Now let me abandon this digression and discuss what maps are valuable in the Pyrenees.

First, upon the Spanish side, there is nothing. Every one who tries to get a good cartographical indication of the approaches to the Pyrenees upon the Spanish side is baffled. Outside of my own experience, I have heard of many attempts and they have all failed. There is indeed a legend of a wonderful military map in Madrid or elsewhere, but I have never seen it, nor have I ever seen any one who has seen it. There is a good contour map extending outwards from Madrid in various sections, but it does not get anywhere near the Pyrenees. There is a geological map of Spain upon which some people fall back in despair, but it tells you very little about Spain except the geology. It is on an extremely large scale, $\frac{1}{400,000}$ if I remember right, and it is horrible to have to use it even for the most general purposes of travel.

There is a large general map of Spain, drawn in Germany, which is equally useless for the pedestrian ; it comprises the whole country within a space that could easily be hung over the chimney-piece of a small room.

In a word, there is no map of Spain for the foot traveller upon the Spanish side. Everything of that kind which exists so far is (I again qualify the statement by adding "to my knowledge") of French workmanship.

It is therefore the French maps which the traveller must consider, and I will detail these in their order with their respective advantages.

It must first be remarked that these maps are to be regarded as official and unofficial ; the official ones should be divided into those proceeding from the French War Office and those proceeding from the French Home Office. The importance of this will appear in a moment.

Of the unofficial maps (which are very numerous) the most important by far is that published and printed by Schrader, and this is important only because it gives contours (at rather large intervals, it is true) on the Spanish side as well as upon the French.

The map can be ordered of Messrs Stanford, and costs twelve shillings for the whole six sheets. Its value consists in giving the traveller details of all the difficult central bit between Sallent and the Encantados. The French contours, as will immediately appear, are easily obtainable elsewhere ;

but to know the Spanish side, the difficulties of the way between Panticosa (for instance) and Bielsa, Schrader's map is a great advantage; it is final on the heights, the steepness, and the changes in direction of the way.

The official maps consist *first* of the War Office maps, the scale of which is $\frac{1}{80,000}$ and $\frac{1}{320,000}$.

The first thing to appreciate with regard to the French maps, is that all of them whether from the Home Office or from the War Office (and in a country such as France the work of these two departments is very different), are based upon the $\frac{1}{80,000}$ survey. It was this survey, undertaken by the General Staff in the course of the nineteenth century, which formed the basis of every other map that Frenchmen use. Certain of its early details were slightly inaccurate, as the heights of the Pelvoux group in Savoy, which Mr Whymper, when he climbed those mountains, corrected. It is, however, the best monument of cartography left by the nineteenth century. Nothing has since appeared to rival it in any country upon the same scale. We must except of course the highly detailed large-scale survey of special districts, which may happen to be, by a political accident, autonomous and wealthy. Belgium has a far better map, upon which indeed all modern work upon the Belgian battlefields is based. Switzerland also has a better map. But no such large area as that of the French Republic has upon so small a scale (much less than one inch to the mile) so complete a record of every track, wood, habitation, height, and water-course.

The $\frac{1}{320,000}$ is merely a reduction of this map; it is of service to people who motor or bicycle, to any one who uses the high road, and who wishes to be able occasionally to wander into by-paths; but for little local details and difficulties it should not be consulted. It is useful advice to any one who desires to know the Pyrenees that he should consult before leaving home a map of the whole range upon the $\frac{1}{320,000}$ scale, but travel in the hills with the $\frac{1}{80,000}$ scale.

The disadvantage, however, of the military map, accurate though it is, and full of detail though it is, lies in two points inseparable from the early conditions under which it was produced; the first of these is the use of one colour, that of printers' ink, so that the line marking a stream, a wall, or a path are similar; the second derives from this, and is the confusion of so many small details, all in *one* colour and in black. There are no contour lines. The hatching, though bold, does not give exact heights, save where such heights are marked in figures, and what with the lines marking the paths in mountainous districts, the water-courses, the roads, the marks indicating the rocks, habitations, etc., the $\frac{1}{80,000}$ map tends (though it still remains the best map for a very careful student, *e.g.*, for a soldier on manœuvres) to be somewhat crowded and confused.

An appreciation of the demerits of these maps, and perhaps a certain rivalry between the two departments, led the French Home Office to undertake an Ordnance Map of its own. This map is in various scales, of which the sheets showing the Pyrenees—the only ones that concern us—are in $\frac{1}{100,000}$ and $\frac{1}{200,000}$. Let me explain the general qualities of both and the advantages and disadvantages attaching to either of these.

Both are in colours, giving water-courses and lakes in blue, woods in green, roads in red, etc., and that is an enormous and immediate simplification upon the old-fashioned black map.

Both are brought up to date with more care than the military map ; both are less crowded with detail, and both indicate such civilian necessities as the telephone, telegraph, post-office, etc. On the other hand, neither contains hatching—the only true way of representing a country side to the eye—and neither gives that minute and exact multiplicity of markings which it is the boast of the military map to afford. The civil map is more practical, the military map more full of duty and more accurate.

It must finally be remembered that the scale of the civil maps, even of $\frac{1}{100,000}$, is so small as to impede the setting down of details such as we have on a one-inch Ordnance Map. It is three to four times smaller superficially than our official map in England. Nevertheless, for reasons that I shall presently show, it is on the whole the best map to carry in the Pyrenees.

The $\frac{1}{200,000}$ map is but a reproduction on a smaller scale of the $\frac{1}{100,000}$ map. It has the great advantage of contour lines, but the scale is so small and the contours so pressed together, that, though it is invaluable for giving a general and plastic impression of the chain (to look down on a general map of the Pyrenees on this scale is like looking down on a model of the French side of the range), it is of little use for telling one, as a contour map should tell one, exactly how much higher this spot is than that other spot. When you are climbing and you wish to identify your position, you have usually to estimate comparative heights on a delicate scale and at a short distance, for which the $\frac{1}{200,000}$ map is of very little use to you.

One way of using the contours of the $\frac{1}{200,000}$ which is laborious, but not without value, is to trace the *deeper* contour lines in some particular district, which you are specially studying. These deeper contour lines stand out much more clearly than the intermediate faint ones, which, as I have said, are too numerous for a mountain district. They can be followed clearly even in the dark shading of a steep ridge, and are set every hundred metres apart. When such a tracing has been made, neglecting the finer intermediate lines, you have a good working relief plan of the mountain you propose to deal with.

Of all the area open to the climber and the man on foot in the Pyrenees, that upon the Spanish side of the frontier is the larger and wilder, and this for two reasons. First, because property and its attendant limitations is more developed upon the northern slope, so that the vast areas common to all, are, if anything, vaster upon the southern side, and secondly, that the formation of the range between the ramparts above the Ebro and the main chain, covers a larger space than that between the main chain and the French plains. Yet, as I have just said, it is on the Spanish side that proper maps are lacking, and one must do the best one can to supplement them by the French extensions.

A common plan guides all the French maps in their delineation of territory south of the frontier. Colours, contour lines, hatching, and every detail are omitted. Heights are given in certain cases (but those are rarer of course than on the French side). The names of towns and, in some cases, their telegraphic and postal communications are marked, but upon the whole the Spanish side upon the French maps has far less detail than is accorded for the territory to which the maps directly relate.

However, let me explain the various advantages and disadvantages, for use upon the Spanish side, of the four types of French maps I have mentioned.

The $\frac{1}{320,000}$ of the Ministry of War may be neglected; whatever use it has upon the French side, it is negligible upon the Spanish.

The $\frac{1}{80,000}$ map of the Ministry of War marks the main water-courses upon the Spanish side, the main peaks, and the main important ports and cols, with their heights, but it does not afford any indication of the shape of the country. It is a bare white space of paper with but few lines traversing it, one or two names, and one or two numbers on each sheet.

On the whole it is better not to use the French military maps for the Spanish side; here it is the maps of the Ministry of the interior which must chiefly be relied upon. Of these the $\frac{1}{100,000}$ map is the best. It is true that the colours, which are so valuable in the differentiation of the French side, are absent upon the Spanish, save in the case of water-courses, which are marked in blue upon either slope of the range. There is no indication of woods upon the Spanish side, as there is upon the French, and as this indication is useful for purposes of camping, the loss of it on the south side is often felt. Moreover, the absence of colour upon the Spanish side often makes one misinterpret the nature of the mountains upon these maps, giving to the whole a bare look, since the rocky and bare spaces on the French side are similarly left uncovered. On the other hand, the $\frac{1}{100,000}$ French map does afford upon the Spanish side a very large number of detailed points of information. I will enumerate them in their order.

1. The general shape of the country is indicated by shading, the light being supposed to come (as is the case throughout this series of maps) from the north-west.

2. Steep rocks and cliffs, the presence of which should always be indicated to the traveller, are carefully marked upon either side of the frontier.

3. Paths, the importance of which the reader will presently appreciate, are clearly marked, with all details, as exactly as on the French side.

4. Every habitation is marked, and in the case of villages and towns, the number of inhabitants, the postal and other facilities.

5. Most of the heights are marked, though not so many as on the northern slope, but at any rate,

the height of every important port, col, and peak appears. In general, it may be said that there is no map of the Pyrenees, immediately to the south of the frontier, equivalent to those of the districts which happen to fall within the French $\frac{1}{100,000}$ survey.

This leads me to the principal drawback connected with the use of the French $\frac{1}{100,000}$ map upon the Spanish side, which is, that it only includes such Spanish territory as accidentally happens to fall within each square blocked out in the French survey.

The English reader is acquainted, it may be presumed, with the one-inch Ordnance Map, and he will have remarked, how, if it so happens that a little corner of land escapes the regular series of rectangles into which the one-inch Ordnance Map is divided, that little corner of land will have a map all to itself, though the greater part of the rectangular space so marked may be taken up by the sea. In the same way any little bit of French territory which projects beyond the scheme of rectangles into which the whole survey is divided, has, added to it, an outer part completing the map and extending into Spain ; where (as for instance on the sheet called "Gavarnie") the little piece of French territory so projecting is small in comparison with the whole rectangle, a considerable piece of Spanish territory will be included ; but where (as for instance on the sheet called "Bayonne") the frontier very nearly corresponds with the survey, very little of the Spanish side will be included.

From this it is easy to perceive that the maximum

amount of Spanish territory in any one map must be inferior either in width or in length to the full dimensions of each sheet, and that the total distance into Spain, which any one sheet can mark, south of the frontier, is less than the width of any one sheet. Now each sheet of the French $\frac{1}{100,000}$ map includes 15 minutes of a degree from north to south, that is, about 17 miles. One may say, therefore, that the amount of Spanish territory shown to the south of the frontier in this excellent survey is always less than one full day's journey. In many parts it narrows to far less than this. There are not a few parts of the range where even for those who make but short excursions on to the Spanish side, this drawback is of considerable effect. For instance, in the easy and pleasant excursion which takes one from Andorra to Urgel, the $\frac{1}{100,000}$ map cuts one short at 42° 30′ below Andorra, and 42° 15′ beyond the main road to Urgel, and no small part of the road lies south or west of this limitation.

The $\frac{1}{200,000}$ map somewhat makes up for the deficiency of the $\frac{1}{100,000}$ map, but not in a complete manner. The frontier sections of this survey (five in number) show Spanish territory to the extent of some 30 miles in the Basque country, they give but a tiny corner of the extreme east of the territory of Aragon, they give over 30 miles for the greater part of the north of that province, but in Catalonia the belt is restricted to far less. Moreover, the Spanish details afforded are much slighter than in the $\frac{1}{100,000}$. There is no indication of the relief of the country, no shading, only the principal water-courses and the principal highways and mule roads are marked. But it is here that the $\frac{1}{200,000}$ is useful, if one has the intention of walking for some days upon the Spanish side. Thus the direction from Castel-Bo in Catalonia to Esterri can be roughly drawn upon the $\frac{1}{200,000}$, and will not be discovered so clearly in any other survey.

It now remains to sum up the respective advantages of these four maps for the general purposes of travel, and to give a few comments upon the uses of each.

The $\frac{1}{320,000}$ military map will not be of great use to the traveller. It can only show him the main roads if he is motoring or cycling, and present him with a general view of the country for which the clearer $\frac{1}{200,000}$ map will serve his purpose better.

The $\frac{1}{80,000}$ military map is the best for minute details, and if a man desires to ramble off and explore some special districts of this great range, it is the $\frac{1}{80,000}$ map which will be of most use to him, though its value will be supplemented and greatly extended by using it in conjunction with the colour $\frac{1}{100,000}$ map of the Ministry of the Interior or Home Office.

This last, as the reader will have seen, is the staple map, upon which every form of travel depends. If no other be purchased, this at least is always indispensable.

It is well here to summarise briefly certain points in the reading of this map, which do not immediately appear on one's first acquaintance with it.

First, the map is on too small a scale to show a

certain number of features which, though unimportant in the general landscape, are essential to
the traveller on foot. This is true of rocks, for
instance ; open rock, extending over a considerable
surface, will always be marked, but hidden ledges,
especially small ones, are more often not marked,
and this may lead to disaster if one trusts the map
too exactly. For instance, in the sheet numbered
xi. 37, a range will be seen rising to the left of the
main road, which bisects the map from north to
south, I mean the range running from the Spanish
frontier to the Pic-du-Ger. This ridge is intersected
by two profound valleys, and the whole of it is a
mass of greater or smaller limestone ledges, more
or less masked in the density of the forests. Yet
it is impossible to indicate these on such a scale,
save here and there by sharp hatching. These
limestone ledges are in this particular case such,
that unless one knows the paths extremely well,
it is impossible to cross the ridge at all, but one
would have no idea of that from merely consulting
the map. On the other hand, every rivulet, however small, is distinctly marked, and that is something of a guide when one has tried to ascertain
one's position in a valley. This map has a further
advantage of marking in the clearest way the paths
by which the various ports are approached, and
after a considerable use of it in many places, I can
say that when you have lost the path, the indication
afforded you by the $\frac{1}{100,000}$ map is invariably right—
upon the French side. However unreasonably the
line seems to acting upon the map, if it lies to the

left of a stream, or beneath a particularly clearly
marked rock, then it is to the left of that stream,
or beneath that rock that you must cast about
if you want to find it, and if you find another
path in another direction, you may be certain it is
but a random track, which will mislead˙ you, however clearly it may appear for the moment. When,
in first using these maps, my companions and I
neglected such information, it invariably led to
trouble. For instance, in the lower crossings of
the Soussóou, the map gives the path everywhere
on the north, or right bank of the stream. There
is a spot just before the first rocky "gate" of this
ravine where all indication of further travel upon
the right bank disappears, and on the contrary
a fine-made path crosses over by a strong bridge
to the further or left bank. We thought the map
must be in error, and crossed by the bridge, with
the result that we spent a whole day cut off by a
bad spate from the further side, and were for some
hours in peril ; for the bridge once crossed, this
false path disappeared within half a mile. If we
had pinned ourselves to the map, kept to the north
bank, and cast about in circles, we should have
found the path again but a hundred yards or so
further on, running precisely as it was indicated on
the survey. The importance of the $\frac{1}{100,000}$ map in
thus giving all tracks accurately will hardly appear
to the reader unused to the Pyrenees, but it will be
seen clearly enough when we come later to speak
of travel upon foot in the mountains.

It is a defect of the $\frac{1}{100,000}$ map that heights, though

accurately marked, cannot always be as accurately
referred to the exact spots standing near the figures.
This is because the heights are marked in pale blue
ink, and the ambiguity is accentuated by that absence
of contour lines which is the chief fault of the series.
The method of marking is to point a small blue
point close to the figures, and this dot marks the
exact spot to which the figures refer. Where the
figures are printed in a white space, and where there
are no other features to interfere with them, this
small blue spot is plain enough, but where they
come upon woodland or steep shading, or other
print, it is almost impossible to discover the dot.
Thus, for instance, in the xi. 37, sheet to which
allusion has just been made, a little lake will be
found right upon latitude 42° 50', just before its
intersection with longitude 2° 40'. The height of
this lake is given as 2170 metres, and the small blue
point to which that altitude exactly refers is
unmistakably marked at the southern extremity of
the lake ; but immediately to the right of those very
figures, one of the highest peaks of the Pyrenees,
the Bat Lactouse, marked 3146 metres, presents no
point of which one can be certain. The frontier
happens to cross this peak, and the little blue spot
has got lost in the chain of black dots marking the
frontier and in the print of the name of the
mountain.

As a general rule, however, if you are in doubt as
to what a figure may refer to, you are pretty safe in
referring it to a peak, rather than to a pass or a
group of houses in the neighbourhood. I have said

that the accuracy of the map is undoubted for the
French side ; it is less certain upon the Spanish,
where indeed its accuracy is not guaranteed. It is
the best map to use upon the Spanish side (save for
that restricted district over which Schrader's contour
map applies), but do not, upon the Spanish side, take
the map against the evidence of your senses, as you
will be wise always to do upon the French side.
The map is notably wrong upon the Spanish side
where unfinished works are concerned ; it is not
revised with the same frequency and care as upon
the French side ; for instance, the big new road
from Sallent up to the French frontier goes in long
winding zigzags, which make the total distance
between eight and nine miles. The $\frac{1}{100,000}$ map
marks it in dots as though it were not finished,
makes it far straighter than it is, and thus reduces
the distance by nearly half.

Finally, the $\frac{1}{200,000}$ map gives the best bird's-eye
view of the whole district, and is the only one
showing contours, and penetrating further upon the
Spanish side than any other. It will be my advice
to those who desire to take a walking tour of
some length in various parts of the range, to equip
themselves with the whole set of the $\frac{1}{200,000}$ maps (5
sheets), with the whole of the $\frac{1}{100,000}$ map, but only
with such of the $\frac{1}{80,000}$ (the uncoloured map of the
Ministry of War) as cover small districts of the
nature of which one is in doubt. Those, on the
other hand, who purpose spending their time in one
or two valleys only, should, without fail, purchase the
sheets of the $\frac{1}{100,000}$ survey covering that district, and

would do very well to add to these all the corresponding sheets of the $\frac{1}{80,000}$ survey.

With these remarks, most that can be usefully told
to my readers with regard to the maps of the Pyrenees
has been told them, but perhaps a few final notes
will not be without their use, thus : The English
traveller must always remember that none of these
maps comes up to the English one-inch Ordnance
for accuracy and detail—the scale forbids this.
Next, let him remember that the dates of revision
of each map will differ, as do the dates of revision of
ordnance maps in every country. For instance, I
have before me, as I write, the $\frac{1}{200,000}$ of Luz, purchased
in this year (1908) ; no date of revision is attached to
it, but the new road (which is at present an excellent
carriage road, one of the best in Europe, up the
Gallego to the French frontier) is marked, at first
as a lane, afterwards as a mule track. On the
$\frac{1}{100,000}$ (Laruns sheet), purchasable this year, the new
road is marked as existing for traffic, but not fully
completed beyond a point about three miles from
the frontier, and its true form is not given but
merely indicated. It is evident that these sheets
were revised at different times (the Laruns sheet
bears a date six years old), and that we must always
take the later of any two impressions, if we can
obtain it. The highways of the Pyrenees upon the
French side especially, both by road and by rail, are
being extended with such rapidity that every year
makes a difference to the accuracy of the information
conveyed.

It remains to enumerate with their titles the

maps covering the district : in England they may be most easily obtained from Messrs Stanford, of 12, 13, and 14 Long Acre, London, W.C. This firm provide the $\frac{1}{200,000}$ for the whole chain of the Pyrenees range mounted on canvas, the most useful map perhaps for motoring and cycling. Any sheet of the $\frac{1}{100,000}$ can also be obtained from them, as all are kept in stock, but by far the most convenient form in which to carry them is to have them folded in the stiff cover issued by the French Government : to get them in this form, a few days' notice in London will be needed. From the same firm the military maps can be procured in a similar manner, but I do not know whether all are kept in stock as a regular thing.

In ordering the sheets of the $\frac{1}{200,000}$ (if one does not purchase them as a whole), reference is made not to numbers, but to names. There are five sheets, " Bayonne," " Tarbes," " Luz," " Foix," and " Perpignan," the price of which in England is 10s. ; the whole series can also be purchased mounted.

The sheets of the $\frac{1}{100,000}$ map may be referred to either by the names of their central towns, or by the index number of the series in which they are printed. It is difficult to say what numbers of these maps exactly cover the range, unless one knows how far from the watershed towards the plain the traveller intends to go. The smallest number sufficient to cover the actual watershed and the highest peaks is 16, or, for the whole frontier, 17. These sheets are by name (going from the Atlantic to the Mediterranean, from west to east), St Jean-de-Luz, Bayonne, St Jean Pied-de-Port, Mauléon, Ste Engrace, Laruns, Luz, Gavarnie, Bagnères-de-Luchon, Val-d'Arouge, St Girons, Mont Rouch, Perles, Ax-les-Thermes, Saillagouse, Ceret, and Banyuls. Referring to their numbers in the series upon the index map, they are respectively viii. 35, ix. 35, ix. 36, x. 36, x. 37, xi. 37, xii. 37, xii. 38, xiii. 37, xiii. 38, xiv. 37, xiv. 38, xv. 38, xvi. 38, xvi. 39, xvii. 39, and xviii. 39. It will be observed that in the index map of the $\frac{1}{100,000}$ series, the divisions running from north to south are marked in Roman numerals, those from east to west in Arabic numerals, and that the gradual increase in Arabic numerals from 35 to 39, corresponds to the gradual trend southward of the Pyrenean chain from the Atlantic to the Mediterranean.

Very few of my readers will be concerned with the main crest of the range alone ; it will therefore be necessary to add to that list northward of the frontier (the lower Arabic numerals) the further sheet according to the district each may have chosen to travel in. A certain number of extra sheets are necessary to those who travel in the main chain only, for instance, " Perles " (xv. 38) includes within the limits of its sheet the frontier upon either side, but this frontier so nearly approaches the northern limits in one spot, that it will be quite impossible to travel in this part until we also add the sheet " Foix " (xv. 37), to the north of it. Even the little lake of Garbet, which is not three miles from the crest of the range, is half out of the map and half in.

Those who desire a complete collection of all the sheets of the $\frac{1}{100,000}$ survey, extending from the furthest mountain over the Spanish side up on the foothills into the French plain, may remark the following lists: in series viii. 35; in series ix. 35 and 36; in series x. 35, 36, and 37; in series xi. 35, 36, and 37; in series xii. 36, 37, and 38; in series xiii. 36, 37, and 38; in series xiv. 37 and 38; in series xv. 37 and 38; in series xvi. 37, 38, and 39; in series xvii. 38 and 39; in series xviii. 39; in all twenty-five sheets will cover the mountainous region in this survey, and any one who desires a complete map of the French Pyrenees, with as much of the Spanish side as the survey includes, should possess them all. The cost of the unmounted sheet in France is 8d., and of the mounted sheet 10d. In England they are sold at 1s. The military map of $\frac{1}{80,000}$ is sold at 1s. a sheet, or 3d. a quarter sheet. The $\frac{1}{200,000}$ is, as I have said, sold in London at 2s., the five sheets 10s.

Schrader's map is in six sheets upon the scale of $\frac{1}{100,000}$ and with contours. It is essentially a climber's map. Detailed maps of special districts of course exist in many shapes, but they must be sought for in the periodical reviews, and in monographs in which they have appeared. Finally, it may interest the reader to know that in the Casino of Bagneres-de-Luchon he may, for 1 fr. 50 c., inspect a fully detailed relief map of the whole range on a scale somewhat larger than one inch to a mile, though the inspection of it rather satisfies curiosity than affords any guide to travel.

Schrader's map is of the greatest value for one particular piece of touring, which I shall describe later in these pages. Meanwhile it may be as well to add a further note upon it here. It is by far the best, so far as it goes, of all the Pyrenean maps; it is due to private enterprise, and if the whole range had been done in the same way there would be no need to discuss any other type, it would amply suffice for all purposes. Unfortunately, whereas the range, within the limits laid down in this book, stretches in length from a degree east of Paris to nearly four degrees west of that meridian, covering, that is, four or five degrees of longitude, and stretches in latitude from 43° 25' to at least 42°, Schrader's survey covers only $1\frac{1}{2}°$ in longitude (namely, from 1° 10' west of Paris to 2° 40'), and in latitude extends over no more than half a degree, namely, from 42° 20' to 42° 50'.

As the reader may see by comparing these bearings with a general map, Schrader's map is intended to include no more than the very high Pyrenean peaks: it is the result of many years of careful individual survey, begun before the war of 1870 and carried on to quite the last few years.

Like the French Home Office map, it is in the scale of $\frac{1}{100,000}$, and, like it, it is printed in colours, but unlike the Home Office map, it shows the invaluable feature *of contours*. You have an exact plan of the country before you, and in clear weather, with the aid of this map, you can fall into no error in connection with the relief of the land. The contours are at some distance, at 100 metres or 328 feet

apart, but this in such country is an advantage ·
indeed, the cramping of the closer contours on the
official $\frac{1}{200,000}$ map, greatly detracts from their useful-
ness. Not only are contours marked, but all rocky
places are given with the greatest care, and the
impression of relief is helped by shading as well as
contour lines. The only drawback of the map,
apart from its restricted area, lies in the absence
of any indication of woods. As to the steepness,
to which woods are often a guide, his contours
amply make up for the deficiency, but for camping
it affords you no indication. On the other hand,
all cabanes and all paths are very clearly marked.

All heights and distances with which you will
have to do in these hills upon either side are marked
in metres, save in the popular talk, which measures
distances by the time taken to traverse them. With
this I shall deal in a moment. Let me first deal
with what is a constant source of trouble to Englishmen on the Continent, the turning of the metrical
system of measure into its English equivalent.

There are two ways of doing this. One is the
application of quite easy and rough rules of thumb,
the other is the more complicated process which
aims at a fairly high degree of accuracy. It is the
first of these of course which most people will want
to know, and there are two simple rules, one for
heights and one for distances.

The rule for heights is, divide by 3, shift the
decimal point one place to the right, and you have
the height in England feet, *within a certain limit of
error*, which I shall presently detail.

The rule of thumb as applied to measures of distance is to take the number of kilometres (a kilometre is 1000 metres, and is, as one may say, the French mile), divide by 8, and multiply by 5, and you have the corresponding number of English miles, *within a certain limit of error*, which I shall describe presently.

For all ordinary purposes these two rules are sufficient, though in both cases they somewhat exaggerate. They make a French distance measured in English miles a little too far, and a French height, measured in English feet, a trifle too high.

The exact constant of error is, in the case of the heights, 1.6 feet in every 100. Thus if your rough calculation gave you a height of 10,160 feet, the exact height ought to be just 10,000; you see upon the map in the blue figures referring to metres, "3048" (which happens by the way to be within two steps of the height of the Bac Lactous). You divide by 3, add a 0, and get 10,160, and you know by the constant of error that the true height is just exactly 10,000 feet.

The knowledge of this constant gives us a rough and ready method of getting a height within a very small degree of accuracy, and for any purposes where such accuracy is required, I recommend it. In consists in cutting off the last three figures, multiplying what is left by 4, and then again by 4, and subtracting that from your first rough calculation. It sounds complicated, but it does not take half a minute, and you will be well within two feet of any

height; for most heights you are likely to calculate, you will be right within a few inches.

For instance, you see 2403 in blue figures upon
the map; dividing by 3 and shifting your decimal
point, you at once get 8010; there is your rough
calculation, which you know to be a trifle in excess
of the truth. Cut off the last three figures and you
have left 8, multiply 8 by 4, and then again by 4,
and you have 128 as the amount of your error. The
peak is by this calculation 7882 feet high, and
rough as the rule is you are within 20 inches of the
truth : the exact height of such a peak in English
feet is 7883.7624.

However, if you want absolute accuracy, multiply
the French measure by 3.2808992, and you will be
sufficiently near the truth to save your soul.

As to distances, the exact proportion of error,
when you turn miles into kilometres by dividing by
eight and multiplying by 5, is 2 inches or so short of
50 feet too much in every mile ; when, therefore, you
are dealing with a hundred miles, you are very nearly,
but not quite a mile out in this form of calculation.
The error is, within a very small fraction, 1%.

If therefore you want an easy rule for turning
your rough calculation into an accurate one, you cut
off the last two figures and subtract from your total
the figures thus left. For instance, 244 kilometres
divided by 8 gives 30½, and that multiplied by 5 is
152.5 ; cut off the 52, leaving " 1 " on the left,
subtract that 1 (making 151.5), and you are within
a few yards of accuracy. As questions of distance
count nothing in mountains compared with questions

of height, I will make no mention of decimals, but proceed to a very different matter, which is the way of counting that the *mountaineers* have, and this you will do well to heed blindly.

When you are tired and distracted and wondering perhaps whether you can push on, if you have the good luck to find a shepherd, he will tell you your distance to such and such a place in *hours*. The Spanish, the Gascon, the Béarnais, and the Catalan dialects all use the same words, so far as sound goes, for this kind of measure, and the Basque will never speak to you in Basque : it is part of the Basque tenacity never to do this. So if you find yourself in any part of the high hills where a man can talk to you of distances, you always hear the same sounds " Dos Oras," " Quart' Ora," " Mi' Ora," and the rest. This habit, as every reader knows, is universal throughout the world wherever true peasants exist ; but in mountains, whether they be Welsh or African, it is not only universal, but it withstands all the invasion of the modern world.

What I would particularly impress upon any one going into the Pyrenees is this, that such a method of counting is exceedingly accurate, and is moreover the only accurate method. Nothing is more fatal to a civilised man of the plains than to take his little measuring stick and measure upon the map by the scale the distance between two points, saying, " It will take me so many hours." There was a Basque at Ste Engrace who very well expressed to me the contempt which mountaineers have for that method of the plains. A deputy of the French

Parliament had stopped in his inn, had thus measured the distance from the village to the pass, and would not believe that it could take three hours. It always takes exactly three hours. I have done it in four by careful dawdling, and the dawdling, when I came to reckon it up, had taken exactly one hour out of the four. Now, measured upon the map, that distance, as the crow flies, is precisely three miles, but it takes three hours none the less. You will not do it in less, and what is odder, you can hardly do it in more, for if you deliberately go too slowly, you are done for in no time, and if you halt, you will find that your halt fits in exactly to make the walking time three hours. Similarly, over the Pourtalet, from the last Spanish hamlet to the first French one, is six hours; part of the way you may choose between a good road and a mule track, but whichever you choose it is six hours; and there is nothing more astonishing in Pyrenean travel than the accuracy of this rough method. As I said just now, you must heed it blindly; it is by far your best guide.

The use of maps has one last thing to be said about it, which applies particularly to the Schrader map and to the $\frac{1}{100,000}$, and this is that where you think you see a short cut, and the map gives you no track, there the short cut is to be avoided. I say it applies particularly to the Schrader and $\frac{1}{100,000}$, because these two maps are so particular in detail that you think their information must be enough without the further aid of a path. Moreover, the path sometimes takes such apparently needless turns that you are for escaping it by an easier cut.

You will never succeed. You may indeed succeed in a bit of exceptionally hard climbing, you may not lose your life, but you will most certainly wish that you had never attempted the unmarked crossing of the ridge you have attacked. It is obvious that the exception to this doctrine would be found in a piece of genuine experiment. If you say to yourself for instance, "I can get over the shoulder of the Pic d'Anie into the valley of the rivulet beyond, which has no name, but which runs into the Tarn of Uterdineta," you will probably do it, but it will not be a short cut from the Val d'Aspe into the valley of Isaba, though it is the shortest way. These temptations for cutting across the hills come very often in one's first experiments in the Pyrenees : they get less frequent as one knows more of them. These mountains are full of vengeance, and hate to be disturbed.

NOTE.—A convenient map for viewing the whole range is the $\frac{1}{400,000}$ which is sold by Messrs Stanford, mounted in two sheets, and in a case. It is especially of use in showing a large belt of the Spanish side. Motorists in particular should see it.

THE ROAD SYSTEM OF THE PYRENEES

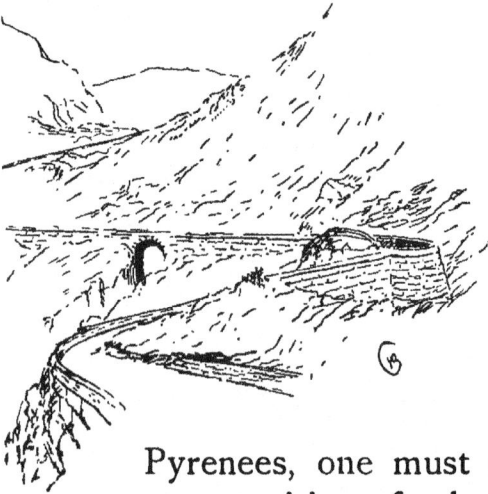

THERE are two kinds of platforms for travel in the Pyrenees, mule tracks and great, highly engineered, modern roads. No others exist. When, therefore, one is describing travel in the Pyrenees, one must separately describe the opportunities of wheeled travel open to all vehicles, however elaborate, and of travel on foot or with a mule. As the last will take up the greater part of my space, I will speak of wheeled travel first.

To understand what are the opportunities of this, one may take as one's standard the roads which can be traversed by a motor car. Those passages which a motor car cannot use cannot be used by a bicycle or a carriage, for the roads of the Pyrenees are, as I have said, either very good broad roads, well graded, and with a hard surface, or they do not exist ; the change is always abrupt throughout the chain from an excellent highway, carefully engineered, to a mule track.

The scheme of Pyrenean roads, as it exists now, is, briefly, *first :* a couple of great lateral roads on the French side, which may be called the upper and the lower road ; *next*, four roads traversing the chain (six if you count the roads along the sea-coast at either end, which I omit—the one goes by St Jean de Luz, the other by the Pass of Lacleuse or La Perthuis) ; *thirdly*, a series of roads, numerous on the French side, rare on the Spanish, which penetrate the valleys but do not cross the chain, and end at a greater or less distance from the watershed.

The main lateral road from the Atlantic to the Mediterranean, along the base of the Pyrenees, links up all the towns upon the plains ; it joins Bayonne to Pau, Pau to Tarbes, Tarbes to St Gaudens, and so on through St Girons, Foix, and Quillan to Perpignan : this may be called *the Lower Road.* The upper road has been but recently completed. It is made up of sections, some of which are old highways, some links quite newly built, and the characteristic of the whole is that it skirts as nearly as possible the crest of the main chain, crossing at some places very high passes over the lateral ridges, and everywhere keeping right up against the high summits of the range. The whole line runs from Perpignan over the Col de la Perche up the Val Carol and over the Puymorens to Ax, Tarascon, and St Girons. At St Girons, it is compelled by the conformation of the country to touch the lower road, but it leaves it at once to pass from Fronsac to

Luchon ; thence through Arreau, Luz, Argelès, Laruns, Oloron, and Mauléon—all the high mountain towns—to St Jean Pied de Port, and thence back again to Bayonne.

The four roads over the ridge into Spain lie all of them on the western side of the hills. They are, first, the road through the Baztan valley, which connects Bayonne with Pamplona ; secondly, the Roman road over Roncesvalles, 12 or 15 miles to the east of this, which used to be the high road between Bayonne and Pamplona before the Baztan road was built, and which was during all history the westernmost road of invasion and communication between Gaul and Spain ; thridly, the road which goes over the Somport, which was also a Roman road and the chief one, uniting Saragossa with the French plains ; fourthly, a road parallel to this and not 10 miles east of it, running over the Tourmalet Pass and joining the Saragossa road lower down. No other roads cross the range from France into Spain until one reaches the Mediterranean, and all these four lie within the first westernmost third of the Pyrenees.

It would be quite easy to open other roads which should unite the last of the Spanish highways with the first of the French, notably over the easy pass of Bonaigo, where 20 miles of work would be enough, and through the Cerdagne, where there are no engineering difficulties. One such road is now in process of completion between Esterri and St Girons over the pass of Salau. Another, which was begun from the valley of the Ariège into Andorra, was abruptly stopped, and it will probably never be

THE GATES OF ANDORRA

completed.   There are some half dozen other places, where a road could cross and where French are building their side of it : but the Spaniards are reluctant to meet them.

Of the roads of the third kind, roads running up the valleys but not attempting to cross the mountains, one may say that on the French side every valley has one or more good roads, the one drawback to the use of which in a motor is that you are compelled, unless you can take a cross road from one high valley to another high valley, to go back by the way you came into the plain.

Not only has every valley its highway leading to the very foot of the main range, but often the bifurcations of the valley will have roads as well. Thus along the valley of the Nive you can go in a motor not only to St Jean Pied de Port, but also right up the eastern valley to a countryside called the " Baigorry " as far as Urepel ; along the next Basque valley to the east, you can go from Mauléon in the plains right up into the hills as far as Larrau, but you cannot go to Ste Engrace, where the valley splits, because the track thither, though a good one, will not take wheels.   You can go up the branch valley from Oloron as high as Aritte, and the main road up the Val d'Aspe (which is that leading to Jaca by the old Roman way), has lateral branches, one taking you to Lourdios, the other across the foot hills to Arudy and the Val d'Ossau.   The valley of Lourdes has a road which, with the exception of the roads over the passes, goes nearest to the main watershed.   I mean the road to Gavarnie ; and the Val d'Aure, which comes next to the westward, has a road going as far as Aragnonette, almost as close to the last cliffs as Gavarnie is; and there is an embranchment to the east which takes one to the very foot at the Hopital of Rivanagon in one of the loneliest parts of the hills. The road to Bagnères de Luchon is carried some miles beyond that town, as far as the Hospitalet, which stands at the foot of the pass into Spain. The road to Viella in the Val d'Oran goes on up to within a mile or two of the pass of Bonaigo. A road from St Girons takes one up the valley of the Lez as far as Sentein, which, like Gavarnie, lies right under the main chain, while the road from the same town up the main valley of the Sallent goes up to the watershed itself, and is being constructed to cross it, and to afford (over the pass of Salau) one more badly needed passage into Spain. The valley of the Ariège has a road all along it, almost to the sources of that river. It is continued through the Cerdagne and down the valley of the Tet into the Roussillon.

There is not a main valley on the French side of the Pyrenees which has not its great carriage road, and most of the lateral valleys have now the same kind of communications. The journey up them is nearly always of the same kind, save the few which are prolonged to carry over the watershed into Spain. There is the succession of two or three enclosed plains or jasses after one has left the plains, the sharp pitch up to one flat, and then another, through short but steep rocky gorges, till we reach the little terminal mountain village, sometimes not more than a group of three or four buildings, lying under the last escarpment, and in sight of the frontier ridge above it. Of this terminal sort was Urdos until Napoleon III. pushed the road out beyond it into Spain; Gabas, until the Republic did the same with the road there; and of this sort still an old Hospitalet, Sentein in the Val d'Aure, and though it is in a state of transition, for the road is now being pushed beyond it, of this sort is Gavernie. Little places almost as old as our race, with no history and no national memories, but with immemorial traditions, rooted as deep as the mountains, were brought into the life of our time by that new activity of the French, which is to many foreigners so hateful, to many others so marvellous.

On the Spanish side there are no roads of this kind penetrating the valleys except the incomplete road to Isaba from Pamplona by way of the Val d'Anso, and the short stretch from Saldinies to Panticosa.

A road is being made up the Val d'Aneu, but it is not yet finished, and a road goes just so far up the broad Segre valley as Seu d'Urgel.

All the other valleys have mule tracks alone.

The general scheme of existing roads in the Pyrenees is roughly as upon the map over page, where it will be seen that much the greater length of the chain is impassable to a wheeled vehicle.

Motoring sets a standard for every other form of wheeled traffic, I will therefore first speak of this kind of travel. The best road to take with a motor, if one wishes to obtain a general idea of the Pyrenees, is the Lower Road (by Tarbes and Foix) from

8

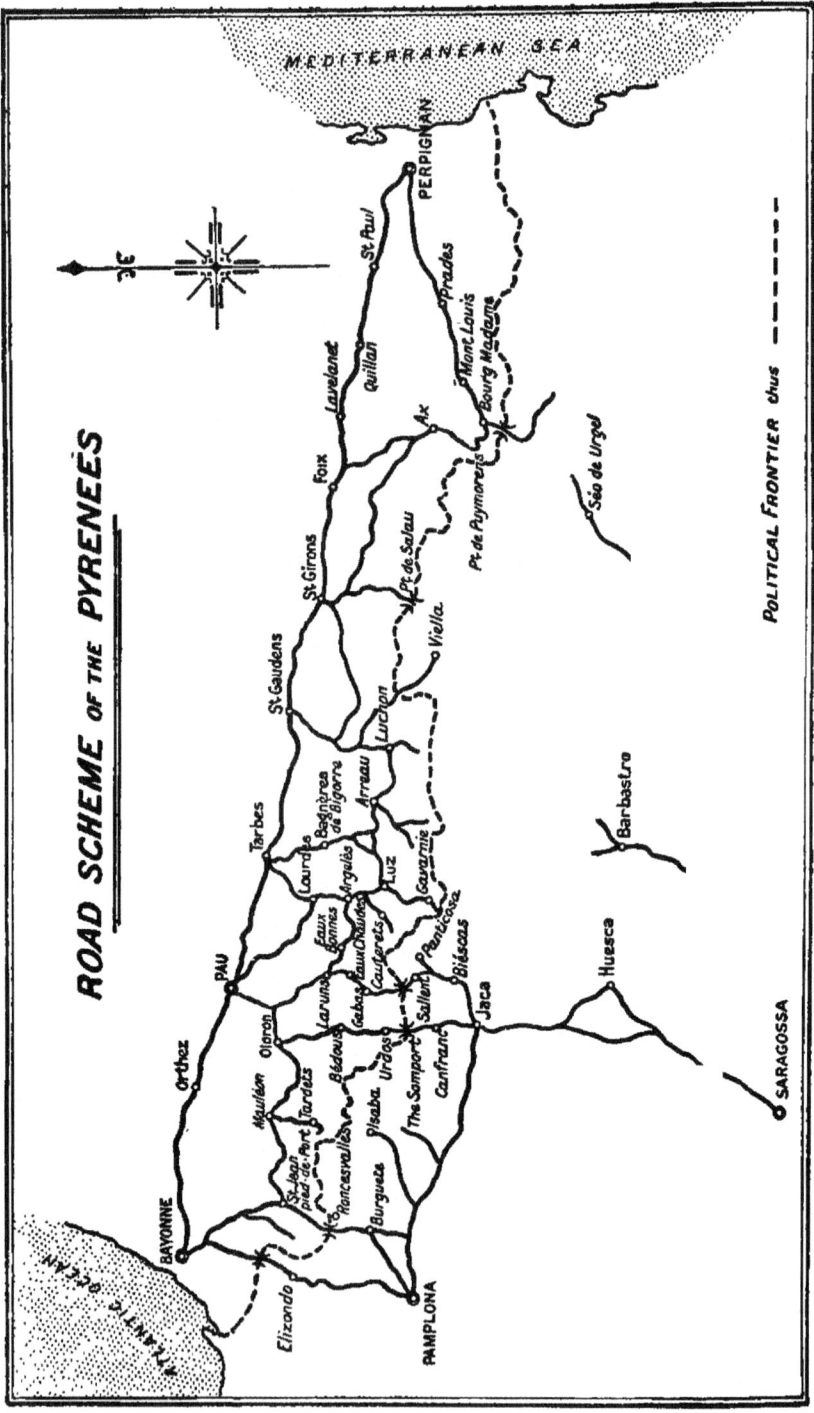

ROAD SCHEME of the PYRENEES

MEDITERRANEAN SEA

ATLANTIC OCEAN

PERPIGNAN
St Paul
Prades
Mont Louis
Bourg Madame
Quillan
Lavelanet
Ax
Foix
Pt. de Puymorens
Séo de Urgel
St Girons
Pt. de Salau
Viella
St Gaudens
Luchon
Tarbes
Bagnères de Bigorre
Arreau
Lourdes
Argelès
Luz
Eaux Bonnes
Gavarnie
Cauterets
Gabas
Pierrefitte
Panticosa
Sallent
Biescas
Larune
Bielsa
Urdos
The Somport
Canfranc
Jaca
Oloron
Bedous
Tardets
Mauléon
Huesca
St Jean pied-de-Port
Roncevalles
Pamplona
Burguete
Elizondo
Bayonne
Orthez
Pau
Barbastro
SARAGOSSA

PLAN L.

POLITICAL FRONTIER thus _ _ _ _ _ _

Bayonne to Perpignan; one may then come back again from Perpignan to Bayonne by the upper road, many parts of which are of very recent construction and which goes right through the highest part of the chain across the main lateral valleys of the Pyrenees. Such a round—about 500 miles altogether—gives one from far and from near the whole of the French Pyrenees: from the first one sees the chain as a whole before one: by the second one mixes with its deepest valleys.

The first day's run from Bayonne had best end at Tarbes; it is a town central with regard to the chain, and it is also a very pleasant place to stop at under any conditions; not cosmopolitan like Pau, and not in a hole and corner like Foix.

The lower road from Bayonne to Tarbes runs through Orthez, Puyoo, and Pau, and if one starts early, Pau is a good halting-place for the middle of the day. This part of the road is, during the whole of its length or nearly the whole of it, a rolling road of the plains with no striking points of view save in where it tops a slight rise. It first follows but runs above and north of the valley of the Adour below it, next descends after the first 20 miles or so to cross the Adour, and so comes to Peyrehorade, the first town (and railway station) upon its course. During all this first part of the run one has sight after sight of the range which stretches out eastward before one to the south rising higher as it goes; and one sees at first before one upon the horizon, later abreast of one and due south, the pyramid of the Pic d'Anie, which is the first of the high peaks.

From Peyrehorade to Pau, between 40 and 50 miles, the road goes through Orthez along the valley of the Gave de Pau, for the most part following the river bank and allowing but few sights of the range ; but at Pau itself it rises on to the high plateau of the town whence the most famous general view of the Pyrenees is spread before one.

From Pau there are two roads to Tarbes ; for curiosity and for general travel it is the road round by Lourdes which is generally taken, and that is during the whole of its length a lowland road though it runs among the foothills ; but the better road on such a drive as I am describing is the direct northern road, which, after it has climbed on to the plateau of Vignan, goes up and down steep small ravines until it comes down again upon the main valley of the Adour and the plain of Tarbes.

There are on this road two points, one just after one leaves the railway line, not quite half-way to Tarbes on the climb up to Vignan, the other just before the loop and descent above Ibos which afford fine views of the range to the south, and one begins to gather one's general impression of these mountains, which, more than any other range, present an appearance of simplicity and the united effect of a barrier. Tarbes, less than 30 miles from Pau, may seem a short run for one day from Bayonne, but it breaks the journey exactly and conveniently.

After Tarbes (where the hotel for you is the Hotel Des Ambassadeurs) the road goes through much broken country, passing by Tournay up on the high plateau of Lannemezan to Montrèjeau. It is a road full of short hills, but it is necessary to take this section in order to go eastward from Montrèjeau and to proceed through St Gaudens, taking an elbow by St Martary and so down to St Girons.

After St Girons one follows the new and excellent road which runs along the valley side by side with the new railway to Foix. From Foix to Nalzen your way is to go along the main road from Foix up the Ariège Valley for some 4 miles and then turn to the left, leaving the railway and making due east. From Nalzen continue to Lavenalet; there take the right hand road to Belesta and Belcaire ; thence, when you have crossed the plateau, a very winding road takes you down, hundreds of feet, on to Quillan. After Quillan you have a few miles through the very little known and wonderful gorges of Pierre Lys to St Martin, through which gorges the railway accompanies you. Do not follow it round by Axat, but cut across by the road which goes eastward to La Pradelle This road takes you across a low pass to the watershed of the Mediterranean. From La Pradelle to Perpignan the road is a perfectly clear one through St. Paul and Estagel. It is a straight, good road, following the valley all the way, save the last stretch, which runs across the plains between the river Agly and the Tet.

This second day will of course be far longer than the first; it is nearer 200 miles than 120. If you would break it, however, break it rather after the short run to St Girons, than at Foix, for though Foix be nearly the half-way house, yet the accommodation is better at St Girons, and so is the cooking.

A two days' run of this kind from the Atlantic to the Mediterranean, following such a route, gives you the whole distant range in one general appearance, and gives it you better than you will have along any other line with which I am acquainted.

The way back by the upper road from east to west through the Pyrenees is a piece of travel quite peculiar to these mountains; nowhere else in Europe is there a lateral road driven right across the buttresses or supports of a main range. The Pyrenees possess such a road in their highest part. What the French have done here is as though the Italians had driven a road from the sources of the Dora Baltea right under Monte Rosa, and the Matterhorn to Lake Maggiore, or as though the Swiss had driven one from Faido and Fusia right over into the valley of Domo d'Ossola. From Tarascon in the valley of the Ariège to Laruns in the Val d'Ossau—that is, over all the central part of the chain and for just over half its length—a mountain road goes right up against the main heights (only once coming near the lowlands at St Girons), crossing the high, perilous passes which lie between the upper valleys. By taking advantage of this new piece of engineering you can return from Perpignan to Bayonne through the midst of those hills which the road just described from Bayonne to Perpignan showed you in a distant general view : when you have so returned you will have seen the heart, the French Pyrenees.

I will now describe such a return journey by the upper road. From Perpignan you will do well to run the first day to Ax. The road is the great road from the Rousillon into France. You go up the valley of the Tet (which is the main River of the Rousillon) through Prades with the Canigou first right in front of you, and at last rising steeply to your left. You continue through Prades up the gorges and tortuous zigzags of the Upper River until you come to the head of the pass at Mont Louis : there the broad and easy valley of the Cerdagne opens to the south, sloping gently before you. The road runs down, almost as in a plain, to Bourg Madame, where you must turn to the right up the Val Carol to Porté. The pass above Porté, (called the "Puymorens") though long, is of an easy gradient, and once over it you run down all the 18 miles to Ax, following the valley of the Ariège.

Ax is, of course, an early stopping place. The whole distance from Perpignan is under 140 miles, but Ax is so much more comfortable than Tarascon that it is better to make one's halt there.

Next day go down the valley as far as Tarascon and there take the mountain road off to the left, it is not a national[1] road but it has a perfectly good surface in spite of a considerable climb. One little col comes almost immediately at Bedeillac, after that you climb steadily up the valley to the Col-du-Port (which is about 4000 feet high) then down the mountain side to Massat, which lies on the western side of the pass and about 2000 feet below it.

---

[1] The French metalled roads are of three main kinds, supported by the State, the County and the Parish respectively. Of these the first and most important are called "National Roads."

Thence it is an ordinary valley road until you come to St Girons again.

From St Girons you continue this progress parallel to the watershed and right among the high peaks, by taking the cross road from St Girons to the valley of the Garonne. Just before the railway station at St Girons turn sharp to your left, taking the road which goes up the left bank of the Lez. At this starting point you are not more than 1300 or 1400 feet above the sea; At Audressein (300 feet up) turn to the right, cross the river, and begin to climb the upper valley until you reach the col of Portet-d'Aspet at about 3400 feet, that is, some 2000 feet above St Girons, and between 15 and 20 miles from that town. From this col the road descends rapidly down the valley of the river Ger, falling in 5 miles 1500 or 1600 feet. At the end of the 5 miles you take a road that goes sharp off to the left before reaching the village of Sengouagnet, this road going off to the left crosses a low watershed, makes, at the end of another 5 miles, a great loop round the forest of Moncaup (the church of which village you leave to the left just before making the turn), and comes down into the great open plain into which the valley of the Garonne here enlarges. It is one of the finest enclosed plains in the Pyrenees, and to come down upon it by this road is perhaps the best way to approach it.

The first village in this plain is Antichan, thence several long windings take one down to Frontignan below, and thence it is a straight road through Fronsac to Chaum where there is a bridge over the river, and where the plain of which I have spoken terminates in a narrow gateway through the hills. You cross the river by this bridge, fall at once into the great national road upon the further or left bank, and a straight run of not more than 12 miles in which one only rises 300 or 400 feet up the tributary valley brings one to Bagnères-de-Luchon. Though at the end of an even shorter day than was Ax from Perpignan, Bagnères will make a convenient stopping-place after a good deal of hill climbing and roads the surface of which, especially in the early summer, is occasionally doubtful. Bagnères has, of course, everything that people motoring can want, it is the capital of the touring Pyrenees, and even if this cross journey has not proved enough for one day, the character of Bagnères make it the right place to stop at on the second day.

Though Bagnères is right in the middle of the mountains, but a mile or two from the frontier of Spain, not 6 miles, as the crow flies, from the watershed and within ten of the highest peaks of the Pyrenees, yet the importance of the town has caused good communications to spring up around it, and there is an excellent road crossing straight over from the high valley of Bagnères into the next high valley, the Val d'Aure. It starts at the market place just opposite the new church, crosses the col called " Port-de-Peyredsourde," and comes down into the main road of the Val d'Aure at Avajan, which follows down the stream at an even gradient to Arreau, 7 miles further on.

Arreau , is the capital of the Val d'Aure, and when you have reached it you will have come about 20 miles from Bagnerès.

The next parallel valley to the Val d'Aure is that of the Gave-de-Pau : the valley which has at its mouth the town of Lourdes, and at its head, right under the Spanish Frontier, the famous village and cliff of Gavarnie   There is, indeed, a small sub-sidiary valley in between where the Adour takes its rise, and of which Bagnères-de-Bigorre is the capital, but it is shorter and stands lower than the two main valleys upon either side.   The section I am about to describe, the great new road from the Val d'Aure to the Valley of Lourdes, just touches this upper valley of the Adour but does not pursue it.

The cross road from Arreau in the Val d'Aure to Luz in the valley of Lourdes is the steepest and the most diverse in gradient, as it is also by far the finest in scenery, of all the new sections which have recently been pierced through the highest parts of the range and between them build up what I have called "The Upper Road."   The distance as the crow flies from Arreau to Luz is not 20 miles, but the long windings of the road which take it over two passes, and the northern diversion necessary to turn the great mountain mass of the Port Bieil, lengthen it to nearly double that distance.

There is no mistaking this road.   It branches off at Arreau, leaving the valley road not half a mile beyond the bridge and going to the left up a little side stream, the name of which I do not know.

Within 2 miles it crosses this stream and begins to take the long complicated and graded turns up the mountain. One must be careful, by the way, at the point where the road crosses the stream to turn sharp to the right and not go straight on towards Aspin, for though one can get to the main road again from Aspin, it is by roads too steep for a motor. If one so turns to the right, the road goes up to the col in great zig-zags and climbs in some 6 or 7 miles the 2000 feet between Arreau and the summit, thence it falls rapidly for 3 or 4 miles to a point where the new road cuts off the corner the old road used to make. It is important to recognise this point, not only because it saves one at least 6 or 7 miles of travelling, but also because it saves one going right down into the valley of the Adour and climbing up again. I will therefore attempt to fix for the traveller the exact place where he must turn off to the left, though the description is difficult on account of the absence of any landmark.

As you come down from the Col d'Aspin, you run through a wood along the mountain side for perhaps 2 miles. The road sweeps round the curve of a gulley on emerging from this wood, crosses the rivulet of that gulley, and comes down close to the stream at the foot of the valley which is the source of the Adour. Just at this point a road will be seen coming in from the left, descending the slope of the valley beyond the stream and crossing it by a bridge. This is *not the road* you are to take. You must continue on the same road you have been following down from the pass, until, in about half a mile, it crosses the stream to the left bank, and approaches on that bank a wood that lies above one on the hill. Immediately after this bridge there is a bifurcation; one branch goes straight on, the other goes off to the left; this last is the one you must follow. The branch going straight on is the old road which leads down the valley of the Adour, and from which one used to have to double back some miles on at an acute angle to reach Luz. The new road, which you must thus take to the left, cuts off that angle.

There are no difficulties from this point onward. The road winds a good deal round the hillside, and almost exactly 5 miles from the point where you turned into it you come again upon the main road to Luz over a bridge that crosses a stream. Just where you join that main road it begins its long climb up to the pass called Col du Tourmalet.

This pass is the highest and steepest on the secondary or lateral passes, over which the new roads have recently been driven. It is just under 7000 feet in height, is everywhere practicable, and once it is surmounted there is a clear run down of some 10 miles and more (following the valley called locally that of the Bastan) to Vielle and to Luz in the main valley.

Of all the crossings between the high valleys of the Pyrenees this is the one best worth taking. The height of the pass, the great mass of the Port Bieil dominating one side of the road, and of the Pic-du-Midi dominating the other, give it an aspect different from any other of the secondary roads, and comparable only to the two main passes of the Somport and the Val d'Ossau.

From Luz a great national road takes one down the valley to Argelès and the railway, a distance of about 18 miles, and the end of about as fine a piece of engineering as there is in Europe. From Argelès, which is just above Lourdes and whence Lourdes can be reached at once by road or by rail, the cross road which I am describing goes on over another high pass into the Val d'Ossau.

The motorist must decide whether to make Argelès his stopping place or not. In distance from Bagnères he will have gone no more than somewhat over 70 miles, and that is a short day; but it is a day that will have included a great deal of climbing and of sharp descents, and that will have had at the end of it one of the highest passes in the Pyrenees. If he does not choose to stop at Argelès, he will find in Eaux Bonnes above the Val d'Ossau, rather more than 20 miles on (but over a high pass), a very wealthy little modern town, like Bagnères on a lesser scale, with everything that he or his machine can want; and only an hour or an hour and a half beyond Eaux Bonnes, by one of the great national roads and along the lowlands, is Pau.

This cross road from Argelès and the valley of Lourdes, into the Val d'Ossau runs as follows. You take at Argelès the road for Aucun, a village about 5 miles off, up a lateral valley, during which 5 miles you climb over 1200 feet.

From Aucun, still climbing, the road passes

Marsous, winds up the hillside away from the stream, and reaches the first pass, the Col de Soulor, thence it makes round the head waters of the Ouzan valley and round the flank of a bare hill called in that countryside "Mount Ugly," until it reaches the point called the Col de Casteix. Here the foot passenger would naturally cross, as he might have crossed still lower down by the Col de Cortes, but for the sake of a gradient the road goes right round to the north and over the Col d'Aubisque, falling from thence in very long curves down to Eaux Bonnes. The town is not 2½ miles from the top of the col in a straight line. It is more than 5 by the long zigzags of the road.

From Eaux Bonnes a road of less than 3 miles takes one down the Pyrenees to Laruns in the valley, and here the great lateral road of the high Pyrenees may be said to end.

One may go to Pau the same night, but, sleeping at Eaux Bonnes, it is a most interesting journey to continue down the valley of the Gave d'Ossau to Arudy and to Oloron, thence by the road through Aramits, and Tardets to Mauléon, thence by Musculdy, Larceveau, and Lacarre to St Jean Pied-de-Port, but all that run is through the foot hills, and though one has fine views of the range from every little pass and hilltop, these last 80 or 100 miles are not of the same nature as the track I have just been describing, the chief feature of which is the presence of a good carriageway running through the very core of high and abrupt mountains. Still, anyone who has taken the lower road, as I have advised, from Bayonne to Perpignan and wishes to go back all the way to Bayonne by a higher road nearer the mountains, cannot do better than go on from Eaux Bonnes to Laruns, to Oloron, Mauléon, St Jean Pied-de-Port, and thence down the lovely valley of the Nive to Bayonne.

So far I have described the main circular journey, west to east, and from east back again to west, which one can take in a motor car in the French Pyrenees.

To describe or to advise as to a similar journey from north to south is not so easy, because the Spanish roads are uncertain. Moreover, there is no Spanish road crossing the lateral ranges as the French one does, so that, unless one abandons the Pyrenees altogether and goes right down into the plains, a circular journey from north to south and back north again is confined to the very narrow choice between Roncesvalles, the Somport, and the new Sallent road.

The road over the Somport is the best International road between France and Spain. Unlike the new Sallent road it is completely finished, and yet it is sufficiently modern to present every advantage for travel. On the French side it has been complete since the time of Napoleon III. ; on the Spanish side its highest stretches have been finished only in recent years. It is perfectly possible to take the whole road from Oloron to Jaca, and so back by Sallent and Laruns to Oloron again in one day, but it would be a foolish thing to do, and if the ascents try the machine, it might mean going through some of the best scenery of the Val d'Ossau

in the dark. It is best therefore, to break the
journey at Jaca, and no number of hours spent in
that delightful town are wasted. The first part of
the road—the first 16 miles or so—are nearly level.
It is interesting to see the straight line which the
Roman track makes for the gate of the hills at
Asasp. The pass seems to invite the road : it is
the most obvious gap in the whole Chain.

The rise, as I have said, is slight. The river,
which is rather less than 800 feet above the sea at
Oloron, is not 1400 above it at Bedous ; in the whole
20 miles or so, you rise but 600 feet. There are
occasional hills, but they are insignificant, and the
general impression is that of following the floor of
the valley. When, however, one has passed through
the great enclosed plain of Bedous, and left behind
him its chief town, Accous, one passes through a
narrow gorge which rises continually to Urdos about
12 miles on. The rise is gradual, however, and
never steep. It was at Urdos that the old valley
road used to stop, until Napoleon III. continued it
to the summit of the pass, and for 7 miles above
Urdos there are continual and steep rises. The pass,
however, is low (it is but slightly over 5000 feet) and
the last 2 miles before the summit are fairly flat.
From the summit the road runs down on the
Spanish side a little steeply, but with no really
difficult gradient, and after about 2 miles of this,
where the Canal Roya falls in and forms the river
Aragon, the road takes on quite an easy slope.
Indeed, the escarpment is so much steeper upon the
French side that Jaca, though it is 25 miles away,

stands no lower than Urdos close by just over the ridge. Rather less than half way between the summit and Jaca is the little town of Canfranc. It would be a pity to stop there, the food is doubtful, and so is the wine, and if one wants to breakfast on the journey, it is better to make an early breakfast at Urdos.

After Canfranc the mountains open out and you are fairly in the lowlands; 17 miles on, through a wide valley, you come to Jaca.

Your Hotel at Jaca will be the hotel Mur, as good and comfortable a one as you will find in northern Spain. From Jaca you may go on to Pamplona westward, or down further south into Spain by Saragossa. As you enter the northern gate of Jaca, you will have gone exactly 57 miles from Oloron; a short distance I know, but I repeat, it is foolish to go to Jaca and not to spend your time in so charming a place. Moreover, the run back has no opportunities for repose.

The return journey is first eastward by the Guasa road, which has (or had, when I went along it last), a most indifferent surface in parts, and you follow this, with a railway never far from the road, some 10 or 12 miles, until at Sabiñanigo the railway turns down south and in much the same neighbourhood (but north of the line) the road turns up north and reaches Biescas (a smaller town than Jaca), in about another 8 miles. After that it begins to climb. At Saldinies the road bifurcates. That on the right goes up to Panticosa; crossing the river by the stone bridge of Escar, your road goes straight

9

on up the valley and climbs up to Sallent for 3 or 4 miles.

I confess I have never been over this bit, but I am assured that it is practicable for a motor, and I have indeed seen a motor which had come round from Panticosa. There is nothing at Sallent that you can call habitable, though as motors live there it is to be presumed that there are ways of looking after them. You will do well to volunteer at the guard room (which is on the left of the road as you leave the town) information as to your whereabouts. It has happened to me not to be allowed to leave a Spanish town without all manner of formalities, while on other occasions it has happened to me to walk through one and over into France without a question being asked.

From Sallent the new road goes up with rather steep gradients at first, zigzaging up the side of the Peña Forata. The old road, a mere track, may be seen cutting off the great bends as one climbs the mountain. About a mile from the frontier, where the steepness of the road grows level, is a post of police where they may or may not bother you; they bothered me on one occasion, and on another they let me alone. From the summit, which is some 12 kilometres and more—say 8 miles by road—from the town of Sallent one goes down first gently, then steeply, with the Pic-du-Midi d'Ossau, a vast isolated rock, right in front of one, and one is accompanied by a torrent upon one's left—which is the Gave d'Ossau. The road follows the right bank of this for some 7 miles, crosses over to the left bank, and 3 miles after this bridge reaches Gabas, a tiny hamlet, where is one of the most delightful hotels in the Pyrenees. Gabas is the highest inhabited point in this valley, and is just the same distance from the summit that Sallent is upon the other side, that is, between 8 and 9 miles. From Gabas down to Laruns the road continues all the way down hill, a matter of another 7 or 8 miles, and from Laruns back to Oloron, through Buzy, is a lowland road with a flat surface. The whole round from Oloron back to Oloron again is somewhere between 125 and 150 miles.

There is but one other circular journey for which I can vouch that it can be made in a motor car ; it is the journey from Bayonne to Pamplona, by way of the low passes on the Atlantic side of the range, and back again through Roncesvalles.

You find yourself at Bayonne as a starting place. The main road into Spain and towards Madrid goes along the sea, much as the railway does, and bears westward, but there is another road through the tangle of Basque mountains, or rather those hills which between them make up French and Spanish Navarre, and this road is the direct road to Pamplona. It is a short day's journey of some 60 miles at the most when all the windings are taken into account, and there are no really high passes or steep gradients throughout. You leave Bayonne by the main straight road which leads out south-west towards Biarritz, but, immediately outside the fortifications, you turn to the left along the high land above the valley of the Nive. A mile and a

half out you cross over the main line and immediately afterwards take the road to the left which
leads you to Arcangues.   There are many branch
roads on this little bit, which is well under 4 miles,
but the chief road is plain.   At Arcangues, just
after you have left the church on the right, you
turn to the left, still following the high road, and
in some 2 miles you strike the forest of Ustaritz,
the confines of which were for so many centuries
the sacred centre of the Basque people.   Through
this forest there is no doubt of the way.   The road
leading to the town of Ustaritz, which goes off on
the left in the midst of the forest, comes in at so
sharp an angle that one would not be tempted to
take it, and the high road goes on, without any
bifurcations, to St Pée.   You have, by this time,
crossed the low watershed between the basin of
the Adour and that of the Nivelle, upon which
river St Pée stands at some 13 or 14 miles from
Bayonne.

You turn to the left in St Pée, by the road that
leaves that village due south, and take the left-hand
road again at the first bifurcation, which is immediately outside the village ; then follow steadily up
the valley of the river.   There is but one doubtful
place, not 3 miles out of St Pèe, where you
choose the left of two roads, but even that is not
really doubtful, for your road obviously follows the
stream, which it there crosses by a bridge, while
the right-hand road goes over into the hills.   About
3 miles more from this bifurcation you cross the
frontier, and thence onwards there is no doubt of

your way. The high road goes over the Pass of Ostondo, or Maya, quite low, and brings you into the Basque valley of Baztan. Come on down through Elizondo, a most delightful town of this people, and climb up continually thence (taking the left-hand road at Irurita, one and a half miles from Elizondo) until you come to yet another pass, called the " Port La Betal " or "Vetale" in French, some 2000 feet or more in height. After crossing this col you are in the basin of the Ebro, and the road thence in to Pamplona is a straight stretch all the way to the plain, which appears suddenly spread out as you round a corner, a fine sight.

The old road back from Pamplona into France over Roncesvalles, the road which the armies of Charlemagne took, and which the Romans built, went first east and west, and was the first portion of the great road to Saragossa. It met the road over the mountains and branched north towards Roncesvalles. There is a modern road which cuts off this corner, and joins the Roncesvalles road quite close to the hills. It crosses three low lateral ranges by very easy gradients, and has an excellent surface. It takes one through Larrasoaña, Erro, and finally, without any doubtful cross roads or turnings, falls into the old Roman road, just below Burguete.

Here you must make ready for one of the greatest sights in Europe. You are on a very high upland plain, something like the glacis of a fortification. The last crest of the Pyrenees stands like a long wall of white cliffs, which seems low and familiar, because you are so very high up on this sloping plain. You go through a fine northern-looking wood which might be in England, with great spacious clumps of beeches and broad glades. You pass the monastery, and then go up through the hamlet of Roncesvalles, quite an insignificant few hundred feet of road ; you see a ruined chapel upon your right (ruined quite recently by fire, and yet no one has taken the trouble to rebuild it !), then suddenly you are at the summit, and a profound trench opens sheer below you and points straight to the French plains, miles and miles away.

It is here that Roland died.

From this summit the road runs down directly on the northern side of the watershed, but still politically in Spain, till you come to the last Spanish town, Val Carlos, where you will do well to ask for papers permitting you to leave the country. These papers are obtained from the Corregidor. Two miles on you cross the river into France, and four miles further you are in St Jean Pied-de-Port, where there is good food and promptitude and news and all that is necessary to man.

From St Jean Pied-de-Port the main valley road takes you, without any doubtful turnings, down the river and the railway, now on one side, now on the other, all the way to Bayonne. There is but one place where the traveller might be a little confused, and that is some 12 miles or more from St Jean Pied-de-Port, where the road, which has been running right along the railway and the river for miles, turns sharp over to the right to reach a village called Louhossoa; but this village (which is but a mile from the river) once reached, everything is plain again. Turn to the left at the church, where the road goes straight back to the river (a matter of 2 miles), crosses it, and goes along the heights on the left bank, all the way back to Bayonne.

The whole of this circle is about equivalent in distance to that which I have described round from Oloron to Jaca, and back again round by Sallent; and, as in the former case, you will do well to break the journey in Spanish territory and at Pamplona, for though this makes two short days in a motor, they are days in which you ought to see what you can see. For my part also, I would stop at Elizondo, to eat and to watch the place; but I would not eat at the hotel in the main street, where the people are cruel and grasping, but rather at the cheap and genial place kept by one Jarégui.

Besides these two circular journeys upon good roads, which a man can take across the main range, there is the variation of them that can be made by taking the valley road from Pamplona to Jaca, a journey of at least 70 miles or more. I know that it can be done, for I have seen motors that had done it, and for all that I know the road may even be excellent: or it may be very bad—I am not acquainted with it. Such as it is, it takes you all along Aragon and the parallel outer ranges of the Spanish Pyrenees.

I have mentioned another extension to the roads described, the run down to Saragossa from Jaca. This of course takes you right out of the Pyrenean country, but the first half of it at least is in the hills, and no journey shows you better the nature of the outlier mountains on the Spanish slope of the main range. Off the direct road one may make a long elbow eastward to reach Huesca, which was St Laurence's town. The surface is good, and there are few steep gradients, though there is a long climb out of Jaca itself. From Jaca to Saragossa, by way of Huesca, along this road, is just about 100 miles, and, as far as Huesca at least, it provides a complete knowledge of the mountain types upon the Spanish side of the watershed. Nor is this typical scenery anywhere finer than in the splendid gorges and chimney rocks of Riglos, nor is any one of the parallel ranges more characteristic than the high Sierra de Guara, which stands up above the burnt plain of Huesca, 30 miles out from the main ridge, quite separate from the general range, and yet reaching a summit of nearly 6000 feet.

All the roads suitable for motoring, especially in such a district as this, are suitable for bicycling also. I say " especially in such a district as this," because the identity between motoring and bicycling roads is more striking in the Pyrenees than in most parts of France, since the expense and difficulty of making the great highways here has been such that it was not worth while building a carriage road on these hills unless the engineering was to be of the most perfect kind, and the surface of the best, and the gradients as easy as nature would allow. The consequence is that there are in the Pyrenees no roads (which he will find in the plains) where a man on a bicycle can go with difficulty, and a motor cannot go at all.   Stretches of this kind, due to bad surface or to steepness, are familiar to every one, but I can remember none of the sort, not even of a few miles, between St Jean Pied-de-Port and Puigcerda, nor between the French plains and the Spanish.

The question will, however, be asked by any one who proposes to bicycle in this district for the first time, whether the long gradients are not such as to destroy the advantage of using the greater part of the roads.   To this objection a general rule applies, one which will seem a little unusual when it is first read, but which I have found from experience to be true.   It is this, that the few crossings of the hills from north to south make easier journeys for the bicyclist than do the lateral roads across the ribs or buttresses of the main chain. Any one going for instance on a bicycle from Laruns to Lourdes, will have some very fine scenery for his pains, and, if the day is fine, he will not regret his experience, but he should be warned that on this lateral road most of his energy will be taken up in slowly climbing the great pass over the Mont Laid ; for though it is but a few miles as the crow flies, it is a big and toilsome business along the highway.   Nor would that be the only pass.   It is characteristic of these lateral roads that they usually contain more than one big ascent.   He will be troubled again at the Col-de-Soulor and to get from Laruns to Lourdes, though the two towns are in contiguous valleys and no further apart than London and Windsor, would be a day's work for most men.

Another example of the same sort could be given from the other lateral roads of the Pyrenees, as, for instance, the low cross road between St Jean Pied-de-Port and the valley of Maulèon. Here the pass is much less high, but a mile or two from St Jean, when you have gone through St Jean-le-Vieux, you begin to climb, and all the long way of the valley of the Bidouze, and out again, over the next range, that overlooks the Saison, is a succession of long wheelings up hill.

For the purpose of seeing some particular place in the next valley, it may be worth while to follow one of these lateral roads, but a general tour of that sort is not worth while. If, on the contrary, a bicyclist chooses the main north and south roads, he will find many advantages in the choice, and I would recommend in particular, as the best that he can undertake in these mountains, the round from Oloron to Jaca and back, which I have already described. Such a journey is a task taking three full days, four or five easy days, and it gives such an opportunity of contrasting two civilisations, and of learning the barrier which separates them, as does not offer itself in so short a space anywhere else, I think, in western Europe. I will not detain the reader in this particular with what I have to say upon this road in general, for that will rather concern the description I will make of it when I speak of travel on foot, but I will point out in what way it can be dealt with by the bicyclist.

All the long road from Oloron to Bedous, though it leads to the very heart of the mountains, needs

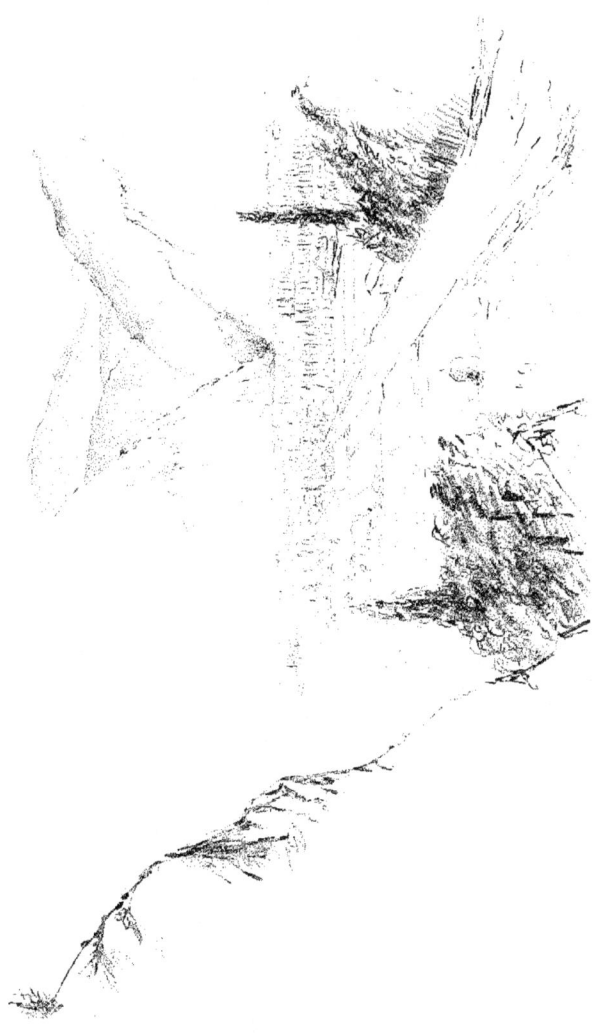

THE ENCLOSED VALLEY OF BEDOUS

no more energy upon a bicycle than does a two-hours' ride (and it ought not to take two hours) in any part of the plains. There are one or two half miles of hill, all of them rideable, but the general run of the way is flat, or burdened with a slight rise which is hardly perceived, and the approach to Bedous, in its magic circle of hills, is actually *down* along a fine slope, which faces the last ridge and the frontier watershed. So far, it is a ride which one may take even upon a high gear, and have for his pains as fine a survey of great mountains as he will find in Europe. From Bedous the road cuts straight across the dead level of the valley floor for $2\frac{1}{2}$ miles, passes a "gate" of rock, and thence continually runs through gorges up the 7 miles to Urdos. It rises considerably in this last bit—nearly 1 in 20—and though the distance from Oloron to Urdos may not take one more than one afternoon, any one bicycling into Spain will do well to pass the night at Urdos, for the big climb begins just after that place. In this hamlet, of no pretensions, you may choose with advantage the little inn called the "Hotel of the Travellers," of which, and whose charming terrace, I speak in another place.

Next day, unless you wish to accomplish a feat, you will begin to walk up to the summit of the road, There are parts that can be ridden—the last quarter is almost flat—but the earlier part and the larger is too steep for comfort. The continental road book makes the whole distance 12 miles, the kilometres by the roadside, which are somewhat more reliable, make it 8, and so does the map; anyhow it is a continuous uphill which should be taken leisurely, pushing one's machine until one gets to the flat bit at the top. The short cuts are here, unlike those of some other cols, quite impossible to a bicycle, even when one is pushing it, and the whole way must be taken upon the high road; if one can afford it, it is wise to have the machine carried on a cart as far as the Hospital, 2 miles from the obelisk which marks the frontier and the summit of the pass; but whether one pushes it, or whether one has it carried, it is a three-hours' climb. It is wisest to take these three hours in the early morning.

From the summit at the entry into Spain there is 2 miles of steep new zigzag, falling a little too sharply, and all around is the very novel aspect of the southern side of the range, where the dryness and the sun have eaten up the forest; at the foot of this zigzag begins an easy and continual run down of 7 or 8 miles into Canfranc; your bicycle takes its own way; there is no place so steep as to fatigue one with the break, still less to be of any danger. The 17 miles from Canfranc onwards towards Jaca is a road upon the whole descending, but by that time one has entered the foothills, which are flat and undulating rather than mountainous, and at Jaca you will find the Hotel Mur, which I have called the kindest little hotel in Europe, and certainly one of the cleanest in Spain.

You will leave Jaca early after spending there your second night. I am not saying that the whole distance from Oloron could not be done in a day, on the contrary, it could be done quite easily. A man could pass the night at Oloron, starting in the early morning from that town, be at Urdos easily by ten, lunch there at leisure, get to the summit by four, and be down at Jaca before dark on a July day, and before the hour of the late Spanish meal. But the climbing of the pass would fatigue him, it would come at an awkward time of the day, and he would have to count upon what is not so certain in the Pyrenees, fine weather. It is best to break the journey at Urdos as I have advised.

From Jaca, a great road leads all the way down to Saragossa, throughout scenery where you are at first amazed by the contours of the isolated cliffs above the gorges of the Gallego, and afterwards almost equally amazed by the aridity of the great plain that slopes down to the Ebro. The run from Jaca to Saragossa is too much for one day in the hot season. It had best be broken at Huesca. If he choose to make this excursion, the traveller will have to return by the same road, and he would perhaps be wise to save himself the tedium of it and to put his machine upon the train, for a railway goes back, much as the road does, to Jaca.

If one does not take the excursion to Saragossa but returns to France, the way is by Biescas, Sallent, and the Val d'Ossau.

The Biescas road leaves Jaca to the east and runs so for 10 miles, then it goes 8 miles northward to Biescas.

From Biescas it begins to rise, in the last part heavily ; and Sallent, which is not ten miles from

Biescas as the crow flies, is nearly 1500 feet higher.
The gorge of approach to Sallent is a plain embranchment from the Panticosa road at Sandinies
about 8 miles from Biescas.

Sallent offers a problem to the bicyclist which it
does not offer to the man with the motor, and that
is the problem of lodging.   It is a bad place to stop
at, and yet the next place where one can sleep is
over the pass, 17 miles on at Gabas.   One will have
gone nearly 40 miles from Jaca, and the last bit one
will have been climbing all the way ; for some miles
up to Sallent quite steeply, and more or less uphill
all the way from Biescas.   To push the machine
up another 8 miles to the summit (for it cannot be
ridden) is a task, but it is a task worth accomplishing,
especially if you have a long evening before you, for
once on the summit you will have not only a run down
of 8 or 9 miles to Gabas without putting your foot to
the pedal, but also the prospect of the best inn in
the Pyrenees, the delightful inn which the Bayous
who own it call the Hôtel des Pyrenèes ; or, if you
like to take the whole pass at once, you have nearly
20 clear miles down hill without stopping, past
Gabas to Laruns ; but the inn at Laruns is not to be
compared with the inn at Gabas.

If one takes on a bicycle the round which I have
spoken of for a motor from Bayonne to Pamplona
by the valley of the Baztan and back again by
Roncesvalles, there is no difficulty about inns, but
on the other hand there is a multitude of shorter
hills, some of which cannot be ridden.   You could
make two short days of the journey out by sleeping

at Elizondo, in which case on your first day you climb up a pass and down into a valley, and your second day is a repetition of the same process. The third day back from Pamplona to France has one hill at Erro, which you will hardly be able to climb, but from that valley through Burguete and right on to the top of the pass is ridable on any reasonable gear. From the summit down to Val Carlos all the way to the frontier is one long easy run down, and you may continue the valley road along the Nive as far as you like upon the same day. Even Bayonne is not too far at a stretch.

As for those who wish to know how to get a series of long coasts in these hills at the least pains, my advice to them is this : start from Perpignan, take the train from Perpignan to Ville-franche ; there put your machine on the omnibus for Mont Louis. From Mont Louis you have a run of 15 miles, falling 1000 feet all through the French Cerdagne to Bourg Madame, uninterrupted save for two or three short rises. At Bourg Madame next day another omnibus (with a very bad tempered driver—at least he was so in my day) will take you up the Val Carol to the summit of the Puymorens ; from there it is an uninterrupted coast all the way down the valley of the Ariège to Ax, and beyond as far as you like to go, 20 or 30 miles of down hill with scarcely an interruption.

The other way round is good coasting too. By the rail to Ax, up the Puymorens by coach, coast down Val Carol, *ride* up (through Llivia) to Mont Louis and coast down the gorges of the Tet. It is only in this eastern part of the range that you will get such long uninterrupted down-hills : there is, in the central part, the run down from the Pourtalet (but no coach to take you up), and there is a coach up the Val d'Aran to Viella, with a run back of a few miles down the Garonne ; but neither of these are like the Ariège valley or that of the Tet, and the roads up the enclosed western valleys to Luz, Bagneres, etc., have not sufficient fall for long coasting.

One ought not to leave the road system of the Pyrenees without saying something on driving. Your best town, I think, for beginning a drive is Oloron, and there is a job-master close to the station from whom you can get horses and carriages by the day, by the week, or by the month. I do not speak of this from my own experience but from what I have been told, and I know that there are relays of horses all up the pass ; but whether the job-master has arrangements for relays I do not know. That sensible kind of travel has so generally died out that I should think it doubtful. It is better to depend upon the same horses for the whole journey, and whether upon the round by Navarre or that by Jaca the posthouses are frequent everywhere, your longest stretches without one being the bit of new road, 17 miles long, between Sallent and Gabas, and the similar 14 or 16 miles between Urdos and Canfranc.

On the other roads, should you determine to drive along them, there is one rather long piece without a relay up the Tourmalet, between the eastern foot of that pass and Barèges ; but this road is continually traversed by carriages at all times, and there is sufficient provision for the distance. These three are the only long gaps without relays which you have to fear in driving through the Pyrenees. For the rest, except that your days' journeys must be so much shorter, what I have said of the roads for motoring applies to driving also.

V

TRAVEL ON FOOT IN THE PYRENEES

THE road system of the Pyrenees and the opportunities it affords for motoring, bicycling, and driving are but a small part of what most English readers desire to know about travel in these mountains. For most men the pleasure of such travel is to be found in wandering upon foot from place to place, in learning a district by slow daily experience, in camping, and in the chance adventures that attach to this kind of life, and also in climbing. Of climbing I can write nothing; it is an amusement or a gamble that I have had no opportunity of enjoying. Those who think of mountains in this way can learn all they need in Mr Spender's book, "The High Pyrenees." They can get more detailed knowledge from Packe—if a copy of the book is still to be bought—and I am told by those who understand such matters that the rock climbing of this range is among the best and the most varied in Europe. In the matter of travel upon foot other than climbing, I have some considerable experience, and this is the sort of travel which I shall presuppose when I come to speak of the various districts into which travel in the Pyrenees may be divided.

There are two ways in which travel on foot in these hills can be enjoyed; the first is by laying down some long line of travel — as over the Somport, across from the Aragon to the Gallego, and so through Sobrarbe to Venasque—the second is by fixing upon a comparàtively small district in which one can slowly shift one's camp from one day to another. In either case, the aspect of travel on foot is much the same, and so are its difficulties and its necessities.

I have heard it discussed whether a man should travel with a mule in these hills. The practice has in its favour the fact that the mountaineers, whenever they have a pack to carry and some distance to go, travel with a beast of burden. The mule goes wherever a man can go, short of sheer climbing, and it will carry provisions for some days. The expense is not heavy; a mule is saleable anywhere in these mountains; one can buy it at the beginning of a holiday and sell it at the end of one, never at a great loss, sometimes at a profit. Nevertheless, upon the whole, the mule is to be avoided. You are somewhat tied by the beast. He is not always reasonable, and feeding him, though it will be easy two days out of three, is sometimes difficult, for while he will carry many days of your provisions, he can carry but few rations of his own. With a mule one always finds one's self trying to make an inn, and that preoccupation is a great drawback to travel in the mountains. Moreover, the keep of a mule, at a Spanish inn especially, is expensive. It is a better plan to hire a mule occasionally, as one needs repose, or in order to carry any considerable weight for a short distance over some high pass.

I presuppose therefore a traveller upon foot carrying his own pack, and I will now lay down certain rules which my experience has taught me to apply to this kind of excursion.

I shall speak later of what sort of kit one should carry, what amount of provision, etc. ; and I shall also speak later of the nature of camping in these hills ; but these two main things do not cover the whole business, and the more you know of the Pyrenees, the more you will find them enemies unless you observe the laws which they teach you in the matter of exploring them.

Now, the first and the most essential of these laws to regulate your travel is to make certain of no one distance in any one time. Do not say to yourself "I will leave Cabanes" (for instance) "and will sleep the night in Serrat." Such plans are too easily made at home or on the plains. One measures the distance upon the map, and the thing seems simple enough. One may be lured into security by starting in fine weather or over easy ground, but *unless you have been over the place before*, never make a plan of this kind, and even if you know the territory, beware of the false confidence which comes so easily in the plains, when one has forgotten the terrors of the high places.

Here are two examples within my own experience

ON THE UPPER STON

to show what dangers attend this sort of confidence, the first taken from the Aston, the next from that very easy place, the Canal Roya; and remember that nothing I am saying has to do with the fantastic exercise of climbing, but only with straightforward walking and scrambling.

A companion and I had settled to force in 36 hours the passage from the Aston valley into Andorra. There is a path marked upon the map; the way is apparently quite clear and one might have made sure that with provision and calculation for one night, nothing could prevent one's reaching the first houses of the Andorrans. On the contrary, this is what happened.

The first evening was mild and beautiful, the sky was clear, the path at first plain. It was so plain that we did not hesitate to continue it after dark. Here was a first mistake, and the breach of a rule I shall insist upon when we come to camping. Still, it was not this error which destroyed us.

We slept the few hours of darkness under a thorn bush before a most indifferent fire, and the next morning we began our way.

We came almost immediately after sunrise to a place where the valley bifurcated, and that in so confused a manner, with so many interlacing streams and so unpronounced a ridge between the main bodies of water, that we took the wrong ascent by the wrong stream, and only found, when we had ended in a precipitous cul-de-sac, that we had made an error. We went back to the bifurcation (which, remember, was of that confused sort where nothing but a very large scale map is of any use), and we made up the other stream. The hours which we had lost had brought us into the heat of the day, and the day was exceptionally hot. We climbed a shelving slope at the end of this further valley : a matter of 2000 feet, very steep and rough. When we were already near the summit there bowled over towards us from beyond it, without the least warning, a violent storm. We were so close to the top, and there was so little shelter on the open rocks we were ascending, that we thought it well to gain the summit before halting. On the whole the decision was wise. We found overhanging ledges upon the summit and took refuge there until the worst of the downpour had ceased. But the storm left behind it a mass of drifting cloud, now rising and now lifting, which made it quite impossible to determine what our true way should be. The summit of the slope was an open grass saddle with great boulders dotted about, and from this saddle a man might go down one of three declivities which branched southward from it. There was no seeing any complete view of the valleys below even in the intervals of the drooping clouds, for, as is so frequently the case in these steep hills, there was a great deal of " dead ground " just below us. We had to guess which of the undulations of the summit we should follow, we could not be certain until we had gone down some hundreds of feet that we had definitely entered an enclosed valley, but once on the floor of this we were fairly certain by our general direction that we had crossed the main watershed and were in Spain.

The storm renewed itself; the late hour made us anxious, we pushed on through the driving mist and rain, necessarily losing a consistent view of the contours and the windings of the valley; when the sky cleared again we saw before us a great open gulf stretching down for miles and miles, and the very amplitude of the prospect further deceived us into believing that we were certainly descending into the first of the Spanish open places, but hour after hour went past and no sign of men appeared. There were not even any huts in the Jasses. To confuse us still further and to lead us on in our error, a definite path suddenly appeared; we naturally made certain that it was the head of the valley road upon the Spanish side. So confident were we that we *must* by the map and by all common sense be now close to habitations that, after consulting together a little, we thought it wiser to eat what little provisions remained so as to gather strength for a last effort, than to camp hungry and reserve our food for the morrow. When we had so eaten it grew dark; hour after hour of the night passed, and the path was still plain—but there was no sign of men. By midnight we were dangerously exhausted and incapable of pushing further : we lay down where we were by the side of a stream and slept. The morning of the third day we might well enough have failed to reach succour. We had come to the end of our powers we had no more food and it was only the accidental encounter with a fisherman who happened to be thus far up in the hills that guided us to safety. He told us that by choosing that particular one of the three slopes we had come down, not upon the Spanish side, but into a long curving valley that had led us back again into French territory. We had made a circle in those forty-eight hours of strain and certainly had we not found him our getting home at all would have been doubtful.

Now these errors for which there seems very little excuse when they are set down thus in print, were not only natural, but as it were, necessary. Anyone unacquainted with the district *might* have made them, and under our circumstances *would* inevitably have made them. Nothing but a large scale map—which does not exist—would have saved us the hours lost at the bifurcation of the streams, and not even a large scale map could have properly decided us at the confused summit of the Pass where a full view, which the storm had prevented, was necessary to judging one's direction. The true remedy lay not in maps, however perfect, but in allowing for the chances of error, in taking a full three days, provision, and in avoiding that sort of forced marching which had exhausted us, and which we had only undertaken from fears about our remaining stock of food.

The other matter, that of the Canal Roya, is the more significant in that it was quite a little detail that might have betrayed us into a very nasty situation. I knew the Canal Roya and, acting on the strength of that knowledge, my companion and I decided late one summer evening not to camp in the valley but to push on over the pass at the head of it, for immediately beyond this pass we knew to lie the good new modern high road which leads down to Sallent. The pass was marked on the map in the clearest possible fashion, the valley was of a very particular and decisive shape, and the pass lay straight over the end of it. Now at that end was a sweep of high land, and rising up from it two rocky peaks. The map and the general trend of the land made it certain that the pass would go to the right or to the left of the lowest of these two rocky peaks. There was no difficulty of approach, and one unacquainted with the Pyrenees might have thought that it mattered little which side of the peak one took, but we both knew enough about the mountains to be sure that there was one way and only one way across; smooth and easy as the approach appeared from our side, all the chances were that somewhere upon the other side there would be precipices. The sun was getting low, and the path which we had been following was suddenly obliterated under a new-fallen mass of scree. Neither of us can to-day ascribe what we did, to anything but luck. We looked at the peak carefully and determined that a certain little notch upon the *right* of it, was the port. We were fatigued after nearly 20 miles of walking (which had already included one Col) and we wearily began the last ascent. It so happened that as we painfully toiled up over and round the loose boulders, the surface to the *left* of the peak became more and more inviting. Our doubts as we surveyed it were like the conflict which goes on in daily life between instinct and reason. Every bit of thought-out reasoning put the port at the little notch on the *right*, but every temptation which could assail two tired men, made us hope and wish against reason that it lay over the smooth grass to the *left*; at last in a cowardly and (as it turned out) salutary moment, we broke for the grass. We tried to persuade ourselves that if that smooth round sward was a cheat, and betrayed (as such enticements often betray in the Pyrenees) nasty limestone cliffs on the further side we had still had daylight and strength enough to come down again and to go up to the rugged notch to which reason and duty pointed. We reached the grass and there found two things, first, the path which had been lost on the stones and the scree suddenly reappeared *there*, and secondly, the descent on the further side towards Sallent was as easy as walking down an English hill.

The reason of this apparent error in the map we soon discovered. Out of sight, beyond the Col, was yet another rocky mass, to the left. The scale of the map was not sufficient to indicate every mass of rock, upon this ridge, but the map, as a fact, did indicate this peak which had been hidden from the valley and was unable specially to indicate the other peak which had been more prominent to us as we walked up from below. The adventure ended well for we got on to the main road before dark and to Sallent before nine, having covered in that accidentally successful day close upon 30 miles. But it might have ended, and should in reason have ended, very differently. For when we looked at the Sallent side of the range the next morning we saw that this notch on which we had first directed ourselves would have led to a perfectly impossible fall of rocks upon the further side. It would have been equally impossible to have gone back in the dark. We should have spent the night on a high stony ledge, without a fire and without shelter and without food, and the next day we should have had no choice but to come down again into the Canal Roya, utterly exhausted, certainly without the strength to climb up again by way of experiment upon other issues, but bound to make our way, if we could, to Canfranc, miles away down the Aragon Valley. It is not certain that we should have had the strength to do this. These examples and many more that one might give, prove the inadvisability of any plan that does not allow for a wide margin of delay : and, as 1 have said, a margin of three days is not too ample.

Not only a misjudgment of topography, to which these hills particularly lend themselves, may put one into a hole of this sort but mist may do it, or worse still, a sprained ankle. Or one may find oneself cut off by marshy ground, or 20 or 30 feet of sheer cliff, too small for the map to mark, may take one an hour out of one's way. In general, allow three days provision for any' task, and never plan single days in the Pyrenees unless you are following a high road.

A second rule is to take the first part of the day slowly and yet without halting. It is the morning usually that gives you your best chance upon the heights, and such examples of mist as have endangered any of my excursions have fallen usually from mid-day onwards. Apart from the danger of mist, if you break the back of the day by ten or eleven, before the first meal, you are safe for the end of it; and breaking the back of the day usually means getting over a port.

A third rule is, stick to the *path*, and if the path seems lost, cast about for it with as much anxiety as you would for a scent.

I have already said in speaking of the use of maps in the Pyrenees, that the great advantage of the $\frac{1}{100,000}$ map was the clear way in which it marked the *paths*. The idea of paths does not fit in very well with the wild life which the Pyrenees promise one as one reads of them at home, and it is of importance to know what a Pyrenean " Path " is, and why such tracks are essential to travel in these mountains.

It is perfectly true that if you are going to camp and fish, or ramble about certain small districts for your pleasure, the point is unimportant, but if you are making a journey from one place to another, upon a set itinerary, a very little experience in the mountains will show you that a " path " must be known and followed, nor do the inhabitants of these hills, whose experience is based upon so many centuries, underestimate the value of these slight and *sometimes imperceptible* tracks. On the contrary, you will hear one of the mountaineers carefully indicating to some fellow of his, who has not yet made a particular crossing, how to find and keep the *path*. You do not hear him giving general indications of scenery, nor distant land marks, but particular directions as to how the path may be made out in passages where it is difficult to trace.

The reason that these tracks are essential to Pyrenean travel lies in that formation of the hills which I have already often mentioned, a formation which causes them to be broken everywhere with sharp descents of rock down which no man can trust himself, and many of which are overhanging precipices. It also lies in the peculiar complexity of the tangled ridges, so that, not even with a good map and a compass can you be certain of guessing your way from one high valley into another.

Now the interest of these paths is that they are not, as the mention of them suggests to one unacquainted with these mountains, definite and continuous. Even the most frequented of them have difficulties of two kinds. The first difficulty is the crossing and multiplicity of tracks as one approaches a pasture, the second is the loss of the way over certain kinds of soil.

Wherever people go to cut wood, or to lead their flocks on to enclosed fields known to them, a divergent path appears and it is often difficult to tell the main path from the branch one. Save over very well-known ports these paths are not made-ways; they are never mended or laid down, they are but the marks left by travel which is sometimes that of but one man on foot in a week, and that man shod in soft and yielding sandals that leave little impress. For many months in the year these faint traces are covered with snow, and in early summer they are soaked in the melting of it. No money is voted for them, and if here and there the crossing a rivulet or the getting past a difficult corner of rock has been artificially strengthened, this will only be upon the main ways and usually only near the villages. A Pyrenean path is the vaguest of things : it is a patch of trodden soil here and there, a few worn surfaces of rock, then perhaps a long stretch with no indication whatsoever. Yet upon this chain of faint indications with only occasional lengths marked, your life depends ; and the finding and picking of it up has the same sort of interest and excitement as the following of a scent or a spoor.

There are three kinds of soil over which the path is almost invariably lost. The first is swampy land, the second is any broad stretch of clean grass, the third is scree.

Loss in swampy land is rare, for the simple reason that the path avoids such land ; loss on scree is often made good toward the end of the summer by the passage of men and animals whose treading down of the loose stones can be noticed from place to place, but intervals of grass are most baffling. The native knows where to pick up the track again upon the further side ; the foreigner has no chance but to guess, from the last direction it took, where he is likely to find it again. He will almost invariably be wrong, and then he must cast about in circles until he finds it upon the further side of the pasture, entering a wood or picking its way between gaps of rock. There is a lacuna of this sort on the perfectly easy way up the Peyréguet, and it cost me last year three valuable hours ; for easy as the Peyréguet is—and it is little more than a plain walk—if you get too much to the right of it, there is a slope on the further side that a goat could not get down.

So much for the importance of *Paths* in the Pyrenees. It is a point very difficult to make in print, but one which the reader, if he intend to walk there, will do well to take on faith. Make the $\frac{1}{100,000}$ map your infallible authority, don't expect to find on the black line it gives—especially if it is a dotted line—more than the merest string of indications, often separated by very wide gaps, and regard the discovery and continuity of these indications as vital to your safety.

I now turn to equipment.

The first question asked by an Englishman about to attempt fresh journeys will be what things he must take with him from England. My answer is. Two things only, his woollen clothing and a pannikin. With regard to this last, the best form is one which I myself get from the Army and Navy stores, and which is of the following character. The handle is double hinged, and curved, so that it fits to the outside curve of the pannikin. A spirit-lamp is sold which just fits into the interior, and with it, a curved metal receptacle for methylated spirit which also fits into the interior. The whole is bound together by a strap, passing through staples upon the sides, and through one upon the cover.

The advantage of carrying this sort of pannikin lies entirely in its compactness. Weight counts. Every ounce counts when you are knocked out upon the third day; and the third day—the forty-eighth hour of losing your way and of missing human succour—may happen to you oftener than you think.

Weight counts even upon the first day, after the first few miles. Weight counts all the time. Now it so happens (why, I cannot tell) that when things are packed in a close compass they weary a man less than when they are loose and straggling, and there is the further recommendation that when they are closely packed, there is less chance of knocking them about and hurting them. So this is the kind of pannikin I recommend. Note, that the people who know most about these hills, the inhabitants of them, carry no provision for cooking. But there is a reason for this which does not apply to the traveller I have in view. The inhabitants of these valleys walk from a house to a house, with the chance of one night at most in the mountains; they carry with them, bread, cold meat and wine, and for the night they make a great fire for warmth but not for cooking. A person exploring at random, and liable to pass several nights in the open, must have the chance of getting a warm meal, and that opportunity will make all the difference if ever he finds himself, as he probably will very frequently, in a tight place. As to the woollen clothing, no one needs to hear the merit of that, and nowhere can it be got so good or so cheap as in England.

Everything upon you should be of wool, except your boots. The differences of temperature are excessive, you are certain to be frequently wet, you will not have a change ; good wool is, moreover, the substance that will wear least in the rough and tumble of your going.

In this connection I must speak of socks. Those who know most about marching, wear none, and for marching along roads it is a sound rule (startling and unusual as that rule may sound) to have the skin of the human foot up against the animal skin of the boot, that boot being well soaked in oil and pliable. There is no form of foot covering within the boot that does not chafe and tear and therefore blister the skin, if one goes a long way at a time, and for many days of continual tramping on end. That is the general rule, and in the French service it is universally recognised in the infantry. Now, to the particular kind of going which these mountains involve that rule does not apply, because, as we will see in a moment, boots are not what one commonly wears. You must therefore take woollen socks—two pairs.

If woollen clothing and the pannikin I have described are to be purchased in England, where are you to get the rest of your kit, and of what kind will it be ?

You must purchase it in any one of the towns of the foothills, and the nearer to the mountains you buy it, the better for you, since the further out you are upon the plains, the more they look upon you, with justice, as a fool who will buy bad or useless material at too dear a rate, and lose, waste, or destroy it in a very few days, a mere tourist to be fleeced.   Buy at St Jean Pied-de-Port, at Tardets (admirable town!), at Bedous, at Laruns (where the people are hard-hearted), at Argelès (where they are too used to tourists), or at Ax.   Buy if you can, *in the fairs;* to these the mountaineers come down to sell their wares and one can bargain, and as for bargaining, I will tell you the prices of things as I proceed.   But of all things do not put off purchasing till you are *deep* in the range.   Do not buy south of Ax, for instance, nor north of Jaca.   The materials grow scanty and bad.

The things you will need are four : first you will need a gourd, next sandals, next a sack, and lastly a blanket.

As to the gourd.   The gourd is the universal vessel used throughout these mountains, and its use extends from an indefinite distance upon the Spanish side (where it is universal) to the towns of the plains upon the French side : to Oloron that is, Mauléon, Foix, St Girons, and the rest. It is a leather bottle of an oval shape, made in all sizes from a quart to a gallon, and this picture represents the structure.   It is in three parts : the oval

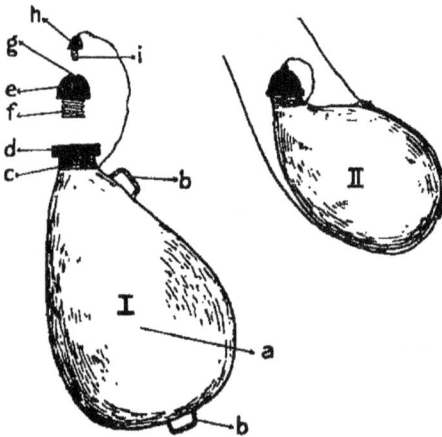

leather case (*a*), which is made of goat's skin with the hair inside; the top (*d*), which is made of goat's horn, with a mouth from an inch to half an inch across, and the nozzle (*e*), which screws on to this top and is pierced by a tiny hole (*g*), through which one drinks, also made of goat's horn. There is a fourth part if you will, the little stopper (*h*), which screws on to the nozzle, and is made of the same material and tied by a string to the mouth of the gourd for fear of losing it. On the inner edge of the leather bottle are two leather loops through which to pass the string, by which the whole thing is carried over the shoulder.

Remember that the name for this invaluable instrument (one has a right to call it invaluable, for it saves the lives of men) is *Gourde* on the French side, and *Bota* upon the Spanish. This detail is not unimportant, for in many French villages they have never heard of a *Bota*, and certainly in no Spanish villages have they ever heard of a *Gourde*. It is in this convenience that one carries one's supply of wine. The horn nozzle on top (*g*) screws off, the wine is poured into the mouth (*d*) through a funnel, until the gourd is completely full; one then screws the top (*g*) on again, and the little stopper (*h*) into that. When one wants the wine to pour into one's mouth or into one's mug, one screws off no more than the little stopper which protects the hole in the nozzle. If you can learn the proper way of drinking out of the small hole pierced in the horn-work, do so. It saves an infinity of delays, and it is the universal method of drinking throughout the Pyrenees. Here is one of those practical things in the trade which you can never get by book learning, and which one can only learn by doing them, nevertheless I will describe it.

Unscrew the little stopper (*h*) and let it hang by its string; take the double horn top piece (*d* and *g*) in the left hand, and grasp with your right the bottom of the leather bottle; tilt the whole up, squeeze slightly with your right hand, held high in the air, and let the thin straight stream of wine from the little hole (*g*) go straight into your open mouth; then (to paraphrase Talleyrand's famous phrase to the Maker of Religions), "if you can possibly manage it," let it go down without swallowing; if you swallow you are lost.

For Talleyrand well said to the Maker of Religions, after having described to him how, to found a religion, he should first suffer obloquy: how he should be ready to stand alone and the rest of it, then added, "If you can possibly manage it," work a few miracles: and this kind of drinking also seems at first miraculous. But it can be accomplished; all it needs is faith, and that strength of will which overcomes the subconscious reactions of the body.

Do not swallow. When you think enough has poured down your throat, do three things all at the same time: relax the pressure of your right hand, tilt the gourd that you are holding upright, and put the forefinger of your left hand smartly down upon the hole in the nozzle. For the first few hundred times you will spill upon yourself a little wine, but in the long run you will learn, and you will drink as neatly and as cleanly as any Basque or Catalan.

If you do not learn to use this instrument thus, you will be compelled to carry a glass, which is not only difficult but dangerous ; and if you compromise by using the gourd, but pouring the wine into a cup, it would either take you infinite time through the nozzle, or else you will have to unscrew the main top piece (e) of the gourd, and if you do that too often it will certainly leak.

These are the elements of the use of the gourd, but, like all things noble, the gourd has many subtleties besides. For instance, it is designed by Heaven to prevent any man abusing God's great gift of wine ; for the goat's hair inside gives to wine so appalling a taste that a man will only take of it exactly what is necessary for his needs. This defect or virtue cannot be wholly avoided, but there is a trick for making it less violent, a trick advisable with an old gourd, when one is starting out on one's journey, and absolutely essential with a new one. This trick consists in pouring into the gourd somewhat over half a pint of brandy and shaking it well up and down, and after that carrying it for a few hours, jolting about and irrigating all the hairy inwards of the bottle as one goes. But do not imagine that the brandy so used can be drunk ; when you have thus used it for a few hours it must all be poured away, for it is wholly spoilt. By the way, if you can get an old gourd second hand that does not leak, it is far preferable to a new one ; all things really worth having are better

old than new.   As to the price of a gourd, you will
not get a small one of a quart or two for less than
2.50 francs, nor a large one from a quarter to a half
gallon or upwards at less than an extra franc for
every quart.   Gourds are not things to haggle
about.   Satisfy yourself that it does not leak and
be grateful to get a sound one.   It will last you all
your life.   As to weight, a gallon is ten pounds:
a quart is two pounds and a half.

Further, you will find very often that when your
gourd is empty, especially if you have carried it
empty upon a cold and misty morning, the inside
sticks together, and when you try to blow it out
through the mouth (as is advisable, before pouring
in the wine), no effort of yours can swell it; the
trick is to put it before a fire and warm it gently;
after it has warmed about ten minutes, it will swell
easily.

As to the sack, nothing is more difficult than
to advise upon this matter.   Some men to be
happy must carry a block, and pencils, and colours,
and brushes.   Others cannot live without combs.
Nothing is really necessary besides bread and meat.
Each traveller must decide his own minimum, but
I can give advice both as to the shape and the
weight of the sack.   The people of the hills,
when they carry a sack, carry a light bag slung
by a strap over the shoulder, and for a light weight,
up, say, to seven or eight pounds, that is the most
practical equipment : thus what we call in England
a satchel, and what the French call a Havresac does
very well.   For anything heavier a knapsack is

often advised; but there are disadvantages in the knapsack: it is complicated, one cannot get at it without taking it off, and it is hot to the back. If you will be at the pains of a knapsack, always have one that is water-tight in material, with a large overhanging flap, and never burden yourself with a knapsack which has outside pockets. The value of a knapsack for heavy carriage is that the weight of it comes right down on to the build of the body. Weight is quite a different thing, when it sags, backward or sideways, from what it is when it presses right down upon the framework of a man's bones. That is why all those used to carrying very heavy weights habitually carry them upon the head or the shoulders, the human body is built for taking a strain in this way down the length of the bones. Now if you carry the haversack by a strap over the shoulder, any appreciable weight, even one so small as ten kilos, becomes a grievous burden after a short distance. Light weights, under that amount, can be so borne, but directly *upon* the shoulders weights up to forty pounds can be carried without destroying a man's marching power, and indeed both French and English armies have often repeatedly climbed the mule tracks of these very hills carrying such weights in this fashion.

It must, however, be remarked in connection with the knapsack that it will not save you fatigue unless the weight bears right down upon the crest of the shoulder blades, and in order to ensure this, make certain of three things. First, that the shoulder straps come well down the knapsack, so that a good part of the weight is above the point where they are sewn on ; secondly, that your knapsack is so packed that the weight is at the top, that no heavy things sag towards the bottom ; and thirdly, that you have strings or straps going from the shoulder straps in front to a belt round your middle, whereby you can brace up the knapsack whenever it begins to lean away backwards. Every soldier knows the difference between a knapsack fitting close to the back and coming well above the shoulder, and one that drags away backwards.

To have said so much about the knapsack may mislead some of my readers. I would not advise it; it is only necessary if for some reason or other you want to carry weight. If you are wise, and content to take only the necessary, a haversack slung at the side from the shoulder will do perfectly well, and it has the advantage of being get-atable at any moment. You may balance the weight of it by carrying the gourd slung over the other shoulder.

As to sandals—Many an Englishman will understand the need of the gourd and the sack who will not understand the advantage of sandals. All the Pyrenean people, for the matter of that, most Spaniards, travel not in leather boots but in cloth slippers with a sole made of twisted cord, and to these the French give the name of sandals. But, as in the case of the gourd, the name suddenly changes on the Spanish side. In France you must ask for *Sandales*, in Spain for a pair of *Alpargatas*. The advantage of these is a thing of which you can never convince a man the first time he attempts these

mountains, but he is sure enough of it at the end of
his first day. For some reason or other, the loose
stones and the pointed rocks of a mule path make
travel upon foot intolerably painful and difficult
if it is too long pursued in ordinary boots. With
Alpargatas on, you do not feel the fatigue of a
track that would finish you in 5 miles if you tried
to do it in leather. And conversely, oddly enough,
a high road with a good surface soon becomes as
intolerable in Alpargatas as is a mule track in
boots. There is nothing for it but to leave your
boots at the nearest town, if you propose to return
to it, or if you do not, to carry them with you and
change from one foot-gear to the other as you pass
from the mountain to the road, and from the road
to the mountains.

Remember that, in Alpargatas, you will *always*
end the day with wet feet. Let not that trouble
you. They dry at once before the camp fire and
they do not shrink. The reason you will always
have wet feet is that in every few miles of hills you
have to cross a marshy place or a stream. But
though it is easy to dry Alpargatas in a few minutes,
it is advisable to change socks at night, while those
you have worn during the day dry before the fire.

As to the blanket—No more than any of the
inhabitants can you go through these hills without
a blanket. It is often of the greatest use in the
changes of weather during the day, it is absolutely
necessary at night. Were you to take it from
England, you would certainly take one that would
be too heavy, or if you took a light one, one that

would be too cold. The people of the Pyrenees who have thought out these things slowly for thousands of years, have ended with the right formula. They have a thin, close, narrow blanket, which just protects a man and protects him as much by its double fold with the air between as by its texture. Get one of a neutral colour, a sort of dark slate gray is the commonest, and pay from 3.50 to 5 francs for it.

With these five things, a pannikin from England, a gourd, a sack, sandals, and a blanket, you are equipped. You cannot take less, you need not take more, and if you take more you will certainly repent it.

I have said nothing about tents. The tent like twenty other luxuries is taken for granted in England. I have heard of people roughing it in various mountains who took with them not only a tent, but an india-rubber bath, a Norwegian kitchen, and for all I know, collars as well. But many a man who will have had the sense to get rid of his luxuries when he begins scrambling, will be reluctant to give up the tent, for it seems necessary to be at least dry. Now the arguments against having a tent have always seemed to me final, so far at least as the Pyrenees were concerned.

You are dealing here with a great expanse of mountain in which weather is very variable, but in which you do not have snow or prolonged furious weather during the months you are likely to travel in. This argument is enforced by the peculiar structure of the mountains. Everywhere in the

Pyrenees you can find either rock shelter—and you find this much more frequently than any other part of the world I have ever seen—or dense forests or, on the bare upland sweeps of grass, those stone cabins of the shepherds, upon the shelter of which the inhabitants largely depend. These, of course, are not very near one to another, but they are always marked on the $\frac{1}{100,000}$ French map, under the title of *Cabanes.* The owners, when they have owners, never mind one's using them, and the only drawback about them is that sometimes you make certain of using one particularly far from mankind, and discover it to be all in ruins. One way with another I have never known three nights upon the Pyrenees which could not be passed in succession without a tent, if the rules which I shall give for camping were properly observed ; and that is the experience also of those who have spent their whole lives in these mountains.

Next, let it be remarked that a tent is a great hindrance, it is either very light—in which case it is always fairly useless—or it is heavy, in which case there is an end to your free going. As will be seen later, when I speak of the way of settling for the night, there need never be occasion for such a shelter, which, moreover, in high winds is more troublesome than an animal or a child.

If your equipment consist in no more than a gourd, pannikin blanket, sack, and sandals, what is your provision to be ?

You must never make your provision for less than forty-eight hours, and it is better to make it for sixty.

However modest is your plan, always allow for two nights on the mountain and for the better part of the third day as well. Remember that you will start in the early morning from the shelter of a roof, that you will therefore have a whole day before you dependent upon your own resources, that if you are making anything of an effort you will certainly camp the first night, but if the weather goes wrong or you miss your way or come upon any accident, you may very well have to spend the second night out, and if you do this, the chances are in favour of a long tramp and scramble on the third day before you reach human beings again. All this will be clearer to the reader when I come to speak of the accidents of weather in these hills, but I may here mention as an example of the truth of what I say that two companions and myself were once held for exactly twenty-four hours in a space of not much more than a square mile, and almost within ear shot of a high road and a village, and that yet it was merely a piece of good luck towards evening—a fog rifting just at the right place for a few moments—that saved us from spending a second night out of doors. In work of this kind the chief part of strategy is to secure your retreat, but you cannot make even one day's excursion without your retreat involving at least another day and perhaps two. Therefore, inconvenient though it be, you must have ample provision.

The first element of this provision is bread, and you will do well to allow a pound and half per man

NEAR THE FONTARGENTE

per day. Those are the rations of the French army and they are wise ones. If each man of a party carries a four pound loaf, you have just enough, but not too much for accidents. A man must have bread, he can do without meat, and at a pinch he can do without wine, but I know by experience that he cannot depend upon any form of concentrated food to take the place of the solid wheaten stuff of Europe. Half a pound of bread and a pint of wine is a meal that will carry one for miles, and nothing can take their place. For meat, you will carry what the French call Saucisson, and the Spaniards, Salpichon. You will soon hate it, even if you do not, as is most likely hate it from the bottom of your heart on the first day, but there is nothing else so compact and useful. It is salt pig and garlic rolled into a tight hard sausage which you may cut into thin slices with a knife, and it is wonderfully sustaining. If you like to carry other meat do so, but you can live on Salpichon and it means less weight than meat in any other form.

These two, bread and saucisson, are the essentials of provision, but other provision hardly less essential should be added to them, and the first of these extras is *Maggi*. Maggi is a sort of concentrated beef essence, sold both in France and in England, and to be got anywhere in the French towns, but you will do well to make quite certain by laying in a good stock of it in some large town, such as Bordeaux or Toulouse or Paris itself, on your way south : I have known the grocers of a Pyrenean town to be out of it. The essence is packed in little oblong capsules which you buy by the dozen, at about a 1d. a capsule, and you will do well to start with three or four dozen a man. They keep indefinitely, they weigh next to nothing, and the great advantage of them will be seen in what follows. You can, with two capsules to a quart of water, make in a few moments a hot and comforting soup which quite doubles the nourishment of your bread, with three capsules to a quart of water you have a very strong soup, which will bring a man round a corner of extreme fatigue. It is a food which can be prepared in a moment under almost any conditions, and one which is invaluable when you find yourself lost, especially if you are cut off by thick weather, or in any other way exhausted. It may seem an insignificant detail to tell the reader how to prepare so simple a meal, nevertheless I will do so. It took me a little time to learn, and he may as well be saved the trouble. Each little cylinder of extract is contained in two gelatine caps which fit together, you pull these off, you drop the essence into a little water while it is warming, but it will not melt of itself, you must crush it and mix it thoroughly with the water, and then add more water, still stirring till you have full measure. It needs no salt in the proportions I have given.

Further, you will do well to fill the little curved receptacle in the pannikin with methylated spirit, and to carry an extra provision of this in your sack. A pint is enough for many days, and very often you have no occasion to use it at all, but you may be caught in some wet place, or in a rocky piece where there is no wood, or in one way or another have a difficulty in making a fire; and even where you have plenty of wood, a drop or two of the methylated spirit makes you certain of the fire catching even in wet weather; of that I shall speak when I come to camping. By the way, take plenty of English matches and of two kinds, fusees and others, and if you are carrying a sack and not a waterproof knapsack, wrap your matches in a little square of india-rubber cloth, for if there is one thing that imperils a man more than another, it is to be caught in the hills without the means of making a fire.

As for brandy, the people of the hills themselves discourage its use; it is, on the whole, best to have some with you, only you must not depend upon it; it is quite honestly, under the circumstances of climbing, what some foolish fanatics think it under all conditions, that is, a medicine. If you take it when you do not need it it will fatigue you, especially in high places. Such as you do take carry in a flask. The gourd, as I have said, spoils it utterly.

Here then you have the rules for equipment and for provision, and I will sum them up before continuing.

For equipment: Haversack or knapsack, a blanket, sandals, a gourd, a pannikin fitted with spirit lamp and spirit vessel, four pounds of bread for each man, a pound of sausage, a pint of methylated spirits, and matches; to which you may add, if you will, a length of candle, and one of those little mica lanterns which fold into the shape of a pocket-book,

and three or four dozen capsules of Maggi. Fill
your gourd with wine as full as it will hold, you will
need it. So much for equipment and provision.

As for the packing of it I have already spoken
of this in connection with the knapsack. A few
additional remarks may be of use. See that your
bread is always covered from the air ; to wrap it in
paper is enough for this, and if it will fit into the
sack so much the better. Work if possible a broad
band of cloth into the straps where they catch the
shoulder, keep the straps short so that the weight
hangs high, carry the blanket loosely over either
shoulder : it gives far less trouble thus carried than
it does when it is rolled and tied over the chest.
If you carry a knapsack, however, roll the blanket
tight upon the top of it, it will then incommode you
even less than when it is carried loosely. Wrap
your matches as I have said in a waterproof cloth
(if you have no knapsack), and wrap in the same
the maps you need for each particular climb ;
forward the rest by post to the town for which you
are making if it is in France ; if it is in Spain, don't,
for they will not get there.

I had forgotten to mention that most useful thing
a pocket compass. Take a large cheap one, and
allow for the variation when you put it on your
map : but of using this and of several other little
points I will speak later. I have dealt with what
regards equipment : let me now speak of Camping.

Camping in the Pyrenees differs from camping
under any other conditions that I know. The
structure of the range, its climate, and even the

political condition of the valleys, make it differ from camping in Ireland or in the Vosges, or in those few parts of England where the wealthy will allow plain men to indulge in this amusement. It is not the same as camping in the Alps, in Savoy, or in the Apennines, or in the Ardennes ; and it is the particular conditions of camping in the Pyrenees which made me say just now that one can do without a tent.

Though geologists are careful to describe the very varied structure of the range, yet to the traveller one feature, peculiar to these among all mountains, perpetually appears common in every part of it, and that is the continual presence of overhanging rock. I can remember no considerable stretch in any main valley, not any in a crossing between two valleys, where you are not perpetually finding examples of this formation. It is this upon which one must first depend for shelter. Next to such overhanging rocks one must depend upon the great forests ; lastly upon the cabanes. But before speaking of their various advantages rules of time must be given, for upon the time of day chosen for the halt the success of a camp will depend.

I am speaking of course throughout these notes of the warm weather alone ; that is, of the end of June, July, August, and the first part of September. Seasons vary, and there are years when the whole of September may be included. At the end of the season one may count, especially in the eastern part of the Pyrenees, upon a sufficient succession of fine nights to make camping possible ; but if one comes upon a streak of bad weather it will last, especially in the western part, for three or four days, and it is better if the people of the valley foresee such weather to let it go over before taking the heights. Thunderstorms and very heavy rain may happen upon any night in these mountains. They are said (I do not know upon what authority) to be commoner upon the French than on the Spanish side. More dangerous than these, though less momentarily annoying, are the mists which gather quite suddenly in the higher parts of the range, and which as suddenly interfere with every form of travel.

It is absolutely necessary, unless one is quite certain of the finest weather, to cross the col or port, in the route one has traced out for the day, before that day is far advanced. The reason for this is two-fold ; first, that wood for a camp fire is not usually to be found upon the higher slopes, secondly that good water is not easily to be found there. It is further necessary to choose the place for one's camp an hour or so before sunset, and it is wiser to make it even earlier. The disappointments which I remember within my own experience in this matter have nearly all proceeded from pushing on from a likely place discovered in the afternoon ; one so pushes on in the hopes of finding a likelier spot before the end of the day. Such an extension of one's journey is nearly always ended in a rough, unsuitable camp, sometimes without a fire, and under the most uncomfortable conditions. When therefore you have found in the course of the afternoon, the shelter of good rock, overhanging a dry place by the stream you are following, pitch upon it and do not regret the hours you appear to lose.

When you have chosen the place for your camp, your first act must be to gather at once as much dry, *large* wood as you can find. The local customs in this matter are very liberal. Even if you are quite close to a village, no one grudges you the use of wood, and your only possible disturbance will come from the frontier guards if you are so foolish as to choose their neighbourhood, which, by the way, can only be the case if you encamp near one of the few chief crossings of the range. These may ask you questions and make trouble, not for your gathering of wood, but for their suspicion that you are smuggling.

The temptation to gather only small wood is strong. It always seems as though the branch you have chosen will be large enough to last for some hours. But a little experience of these fires will show you that nothing small enough for you to drag will be too large for your purpose. The eight hours or more during which you must feed the fire consume a great deal of wood, and the keeping of the fire in depends upon having large logs for its foundation. You will not, of course, be able to cut these into the right length, you will have so to arrange them when the fire is once well started that they burn through their middles. You can then, later, shift into the centre of the flame the halves that fall aside. If there is any breeze pile a few stones to windward of your hearth, for you will have to sleep to leeward of the fire, and an arrangement of this kind will break the force of the wind and prevent the smoke and flame from coming too near you. If the wind is too strong, you must make your fire and your camp under the lee of some great rock, or it will both burn out in a very short time and make itself intolerable to those who depend upon it for warmth. For a wind that rises in the middle of the night, you have, of course, no remedy ; short of heavy rain it is the worst accident that can befall you. If you have enough wood make your fire of a crescent shape with the hollow towards the wind. It is the warmest and the best way. You must so arrange that in sleeping you lie with your feet towards the fire, and your great provision of wood must be brought quite close to hand, otherwise most certainly, you will not have the energy to feed it in the few wakeful moments of the night. That wood should be somewhat green or wet matters little if you have a great fire well started, but if you let it get low while you sleep, it will be impossible to revive it, and when the fire fails, there is an end to sleep for everyone. It is impossible to say what the effect of such a fire is by giving reasons for it ; it does not perhaps warm one so much as do something to the air which makes sleep possible and easy without a shelter, and it is the universal aid and solace of all the Pyrenean mountaineers, whom you will often find in groups, woodcutters or shepherds, gathered round one of these great blazes for the night.

The conditions of a good rock shelter, of a neighbouring stream and plenty of wood, though common, are not universal, and if from the structure of the hills and from the nature of the map you fear you will not reach one, or if the greater part of the afternoon is passed without your finding such a place, your next choice must be a spot in one of the great woods that everywhere clothe the range. They are more common upon the French than upon the Spanish slope. Here there is always cover from the wind, for they are very dense, and even a partial cover from the rain; but it is important to make your fire in a clearing, and luckily there is nearly always a succession of open spaces between the forest and the stream. With such a fire and with such an arrangement to leeward of it the Pyrenean blanket with which you have provided yourself will be ample covering for the night.

As for using cabanes, I have already said that there is no grudge felt against you for doing so, but you must treat any man coming upon you in such a shelter as though he were the owner, for the local shepherds will certainly regard you as their guest, and will think they are doing you the favour of a host. Moreover your fire, if you make one here, must be lit outside the building, though the local people who use the cabanes most constantly, will often make it inside. On the whole the night is more comfortably spent in the open than in one of these shelters, unless one is caught by rain.

The open sandy spaces such as are quite common by the side of the larger streams may be used with safety. There are no places where a spate will be so rapid as to endanger one, unless one choose as

a companion and I were once compelled to choose,
a cave almost cut off by the water.   The only places
where it is essential that one should *not* camp, are
the higher flats where wood is rare, and where the
cold of the night is exceptionally severe.   It is a
choice to which one is often compelled, if one pushes
on too long, after having miscalculated the fatigues
and duration of the climb;
but it is an error which
one always regrets.

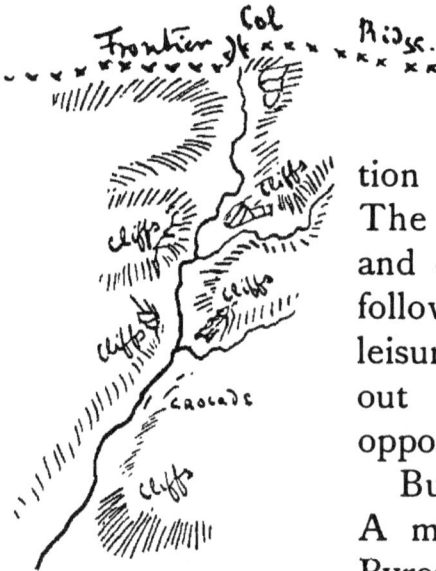

A further recommenda-
tion is, *not to camp by the map.*
The map may look like this
and one may say that one will
follow up the stream at one's
leisure.   The reality may turn
out like the picture on the
opposite page.

But it is a great temptation.
A man may have known the
Pyrenees and experienced time
and again the error of trusting to a map for a camping
site, but there is something so convincing about the
print and the colours that after years of experience
one may commit the same folly again.   It was but
this year that, trusting to the $\frac{1}{100,000}$ map, I planned to
camp at the place where the Cacouette falls into the
main stream below Sainte Engrace.   I did not know
the spot ; it seemed to come at a convenient hour in
the ascent of the mountains : I should be there about
5 o'clock.   There was wood marked, good water ;
it was on the lee side of the wind that was then

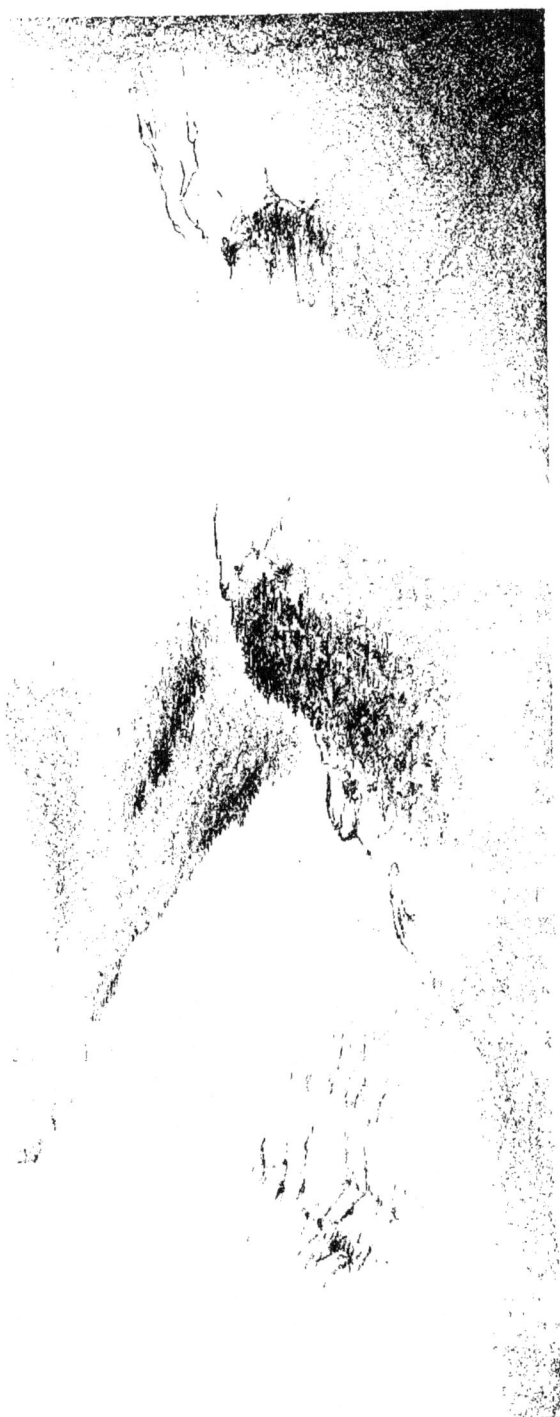

blowing from the south. When I came to it the place was a sharp ledge of limestone higher than Cheddar cliffs, dotted here and there with trees and affording between the wall of rock and the water not three feet of ground. It was not to be approached from above ; it could not be reached from below. A more impossible place for camping never was. I had the same experience some years ago on the Aston, though that was before I knew the Pyrenees well. There a place chosen by my companion and myself for its mixture of wood and meadow upon the map, there were cabanes and apparently plenty of good water ; it was so plain on the map, that one did not hurry to reach it before darkness ; but when we got there it was a marsh ; no cabane appeared until daylight, and there was even that very rare thing in the Pyrenees, doubtful water. As for the wood that should have dotted the pasture, it turned out to be tough little live bushes, and all green, that would neither cut nor burn.

There is one last and very grave danger of which I would warn the reader in connection with travel on foot in the Pyrenees, with a map and even with a map and a compass. Without map or compass it is more than a danger, it is a sort of necessary misfortune perpetually attending men, and the gravity of it is proved by the fact that the local people who use neither compass nor map when they go into a district with which they are unacquainted, carefully ask the marks of the path and get themselves accompanied, if they can, by someone who knows the countryside. This danger may be called " Getting into the wrong valley."

As one sits at home, one thinks of the scheme of mountain valleys too simply. One thinks of the stream as coming down through a ravine with its head waters appearing below a definite saddle or notch in the watershed. This stream, let us say, is flowing north. One sees on the further side another stream rising just on the other side of the notch and flowing on through a simple valley, going to the south. The crossing of the port between these valleys seems to depend upon no more than physical endurance and fine weather. One goes up one stream to the saddle, crosses the saddle, follows the other stream down hill, and so makes one's passage from France to Spain.

There are many passes of this simplicity, but there are many more that, both between the lateral valleys and over the main range, present the danger of which I speak, and which consists in a complexity at a summit such that it is difficult in the extreme to know—even when one is certain one has gone up the right part of the hither slope—what one should do on the thither.

This danger of "getting into the wrong valley " cannot be seized without illustration, and in the following rough sketches I give examples of this.

Here, in the first example, is a bit of country such as one very often gets in these mountains with summits round about the 2600 metre line and the last valleys under the ports somewhat above the 2000. I have marked with hatching the contours

below 2200 and in black the summits above 2600.
The main watershed I have indicated by a dotted
line.

When one is crossing a port of this type one

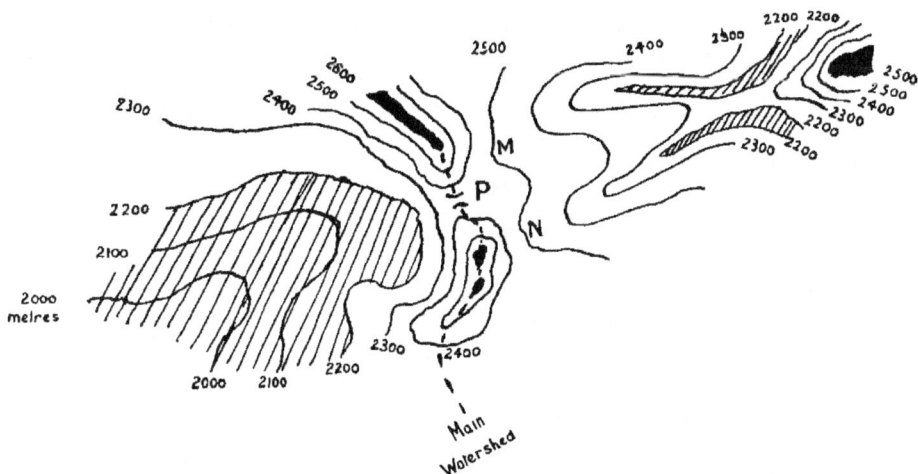

sees before one from the summit a confused and
gentle slope leading apparently to one obvious
valley on the far side like the obvious valley out of
which one has just ascended.   It seems indifferent
whether one should come down on to this by M or
by N, to the left or to the right, yet the two valley

What a man sees from the summit of the Port at P.

floors to which each leads are quite separate and
may lead one round to different river basins.   How
deceptive such a place is, the rough sketch appended
may help the reader to grasp.   It shows the kind of

thing one sees from the summit of such a pass and how indifferent the choice appears between the ways by which one may descend.

This type of confusion exists sometimes in a still more dangerous form, as in the contour lines of this second sketch.

A man arrived at the port P climbing up from the

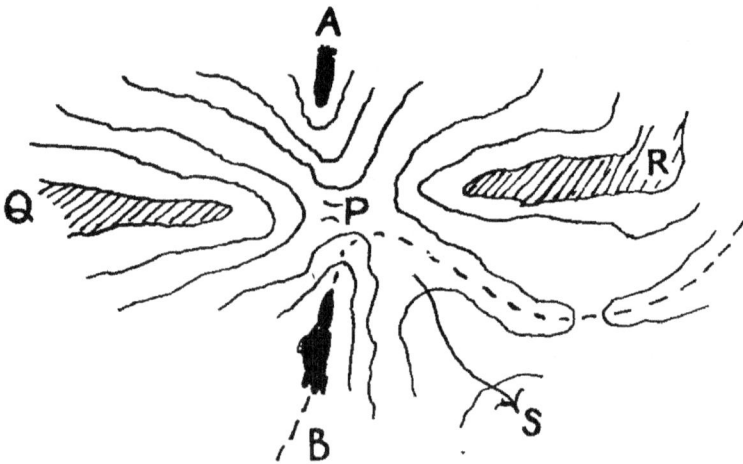

Main Watershed.

valley Q, which is deep and well defined, sees before him another valley R exactly in line with the last, also deep and also well defined. On either side of him, as he gets to the saddle, run high ridges perpendicular to the line of the two valleys. It seems common sense to take the watershed as running along these ridges and across the port, and if Q is the French valley, R will be the Spanish one. As a matter of fact the watershed may not run in this simple way at all, but (as indicated upon the sketch map) take a sharp turn to the right. R may be a French valley after all, and the proper way down into Spain may be over the gradual grassy slopes indicated by the arrow line.   A man standing

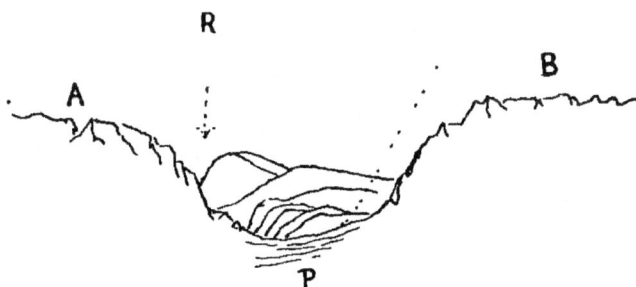

What a man sees just on arriving at P.

just at the port, and having a rocky ridge A and the rocky ridge B to his left and right, sees before him the obvious trench of the valley R and takes for granted that it is the Spanish valley, whereas his true way is across the vague grassy land towards S, and the watershed which he thinks runs from B on to A really turns round from B and runs on to the distant mountains before him.

It must be remembered that on these summits all traces of a path as a rule disappear.   What is worse, indications of a path may begin on the other side into the wrong valley and not into the right one.

A second type of this peril is that in which some feature upon the ridge which looks quite unimportant upon the one side turns out to be all-important upon the other.   Thus a man coming from A in the map over page, where the valleys are hatched and the highest summits are black, would have before him the plain ridge B C.   It is indifferent where

he crosses it from that side, but on the far
side he finds a confusion of falling valleys, and if

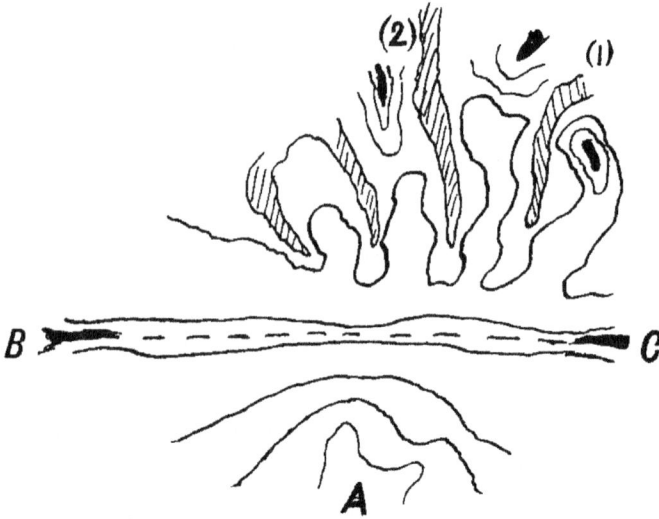

he does not pick out the right one he may find
himself in a few hours shut in by high walls which
constrain him to a journey he never meant to
make. He may have intended to follow valley
(1), and so to reach food and shelter, he may find
himself in valley (2) caught for the night far from
men and with walls of 3000 feet between him and
them.

Sometimes this confusion takes the form of one's
being led on to an obvious notch in the ridge before
one : a notch lower than the general line of the ridge
which (one thinks) cannot but be the port. When
one has climbed to it, however, one finds that the
valley one was seeking lies far to the right or to the
left of such a notch, and that the gap which was so
noticeable on the one side of the pass corresponded
to nothing useful upon the further side.

There is a good example of this under the peak called Negras where an obvious notch which one thinks surely must be the way over to the Gallego, leads to nothing more useful than an enclosed Tarn under the precipices of the mountains.

A sketch of the aspect of this particular ridge will make the difficulty plain.

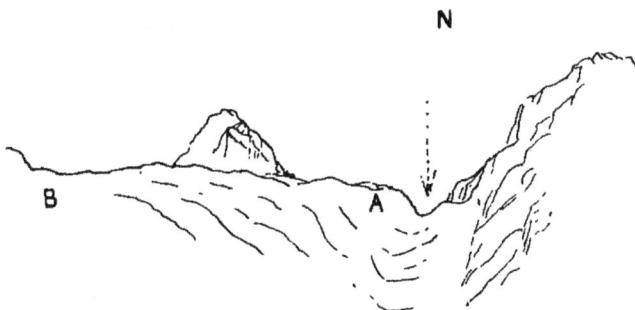

All the contours upon the Aragonese side invite one to the notch at N, yet the true way lies over the ridge between A and B, and the nearer to B the better is the descent upon the further side. Indeed at A it is perilous, at B it is a very gradual descent of easy grass.

The third type of mountain structure which may lead one into the wrong valley is what may be called " The Double Col." It is damnably common and a good example of it will be found in the track I describe later on in this book when I speak of the short cut from the Ariège Valley into the Rousillon.

The accompanying sketch will explain the character of this sort of tangle, and it is most important that anyone unacquainted with these mountains and wishing to learn them should seize it thoroughly, for it is the worst of all the lures that get a man astray.

Observe carefully the numerous contours upon this little sketch map. They are numerous because

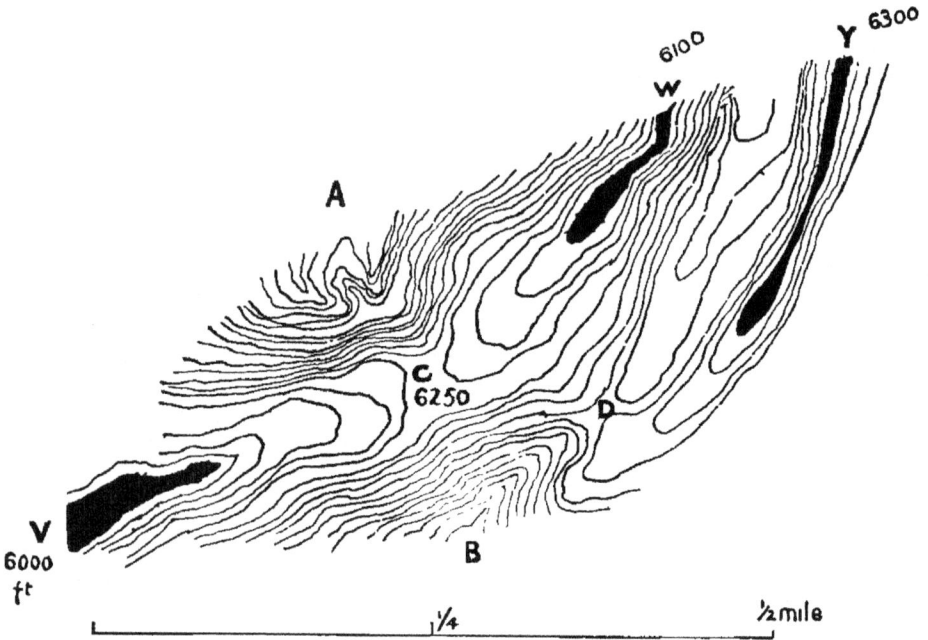

it is necessary to show the minute details of such a case. I will suppose them to be about 50 feet apart. The traveller is coming up the valley marked V, the floor of which is marked in black upon the sketch, and the apex of which is, let us say, 6000 feet above the sea ; he climbs the last little slope of 250 feet and reaches the col at C, which is 6250 feet above the sea. On this saddle he has upon either side of him precipitous slopes, which lead up to two summits of mountains upon the right and the left, the one towards A, the other towards B. Right in front of him opens another valley corre-

sponding apparently to the valley V from which he
has come, and which we will call W. The floor of
this also is marked in black upon the sketch. It
will be observed from the contour lines that the
descent on to W is easy, though the walls bounding
it on either side become increasingly precipitous as
one proceeds.

Hidden from him by rising ground upon the
right, as he stands at C, there is yet another valley,
the floor of which is also given in black. This
valley we will call Y, and it is this valley which
leads the traveller towards his object ; valley W
only gets him deeper into the wilderness. Both
valleys W and Y, are so precipitous that once en-
gaged in either of them one is caught and compelled
to pursue them for many miles. It is evident that
on a very large scale map such as this, and with full
contour lines giving every few feet of height, the
traveller would make no error. Once at C he would
go up to the right around the base of mountain B,
rising continually until, somewhat under 6500 feet,
he came to the second col, D, which would bring
him down into valley Y.

But consider how this corner would look upon an
ordinary small scale map !

The whole distance from the apex of valley V to
the apex of valley Y is not half a mile. It would
occupy little more than a quarter of an inch upon
your French map. The general trend and nature
of the valleys, which the traveller shut in by high
mountains cannot grasp, would seem obvious upon
such a map and he would take it for granted that he

could make no error and that the passage marked from
V to Y would be perfectly plain sailing.   It would
never occur to him that he could be trapped into the
little ravine W leading nowhere and in no way connected with his journey.

The map would look something like this, perhaps,
giving one a perfectly accurate general impression

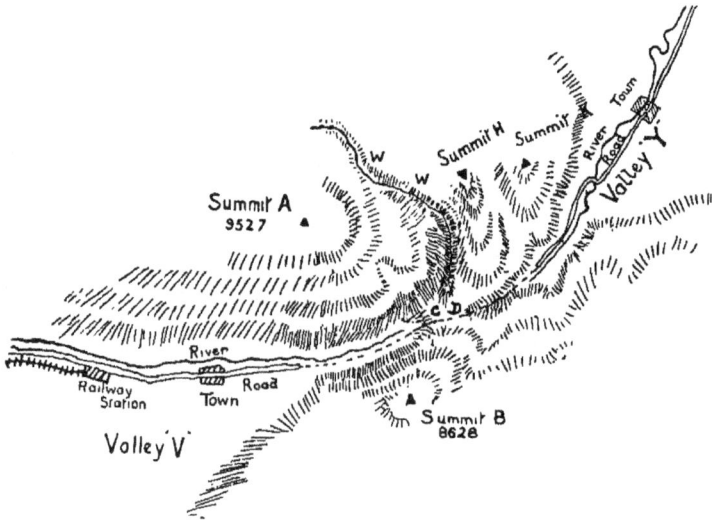

of the whole country side, but quite useless for the
critical point C-D, the difficulties of which nothing
but numerous contours and a very large scale can
possibly explain.   The traveller consults the map,
he sees the mountain group whose summits are A,
H, and K, with their heights marked, he sees the
other mountain group culminating at B with its
height also marked, he sees the main valley V up
the road of which he has proceeded with the town
in which he stopped and the river which he has been
following.   He sees the pass clearly marked at C-D,
leading over to the further valley Y with its town,

river, and road—and the journey seems to present no difficulties.   It is only when he gets actually shut up in the hills at the heads of the valleys that he may begin to doubt or to be misled.   On his map he could never believe that the little torrent W going right round out of his direction could take him in, or that he would get into its valley.

If you consider what he actually sees when he gets to the summit of the pass, you will appreciate yet more easily how his error will come about.   He will see something like this, with an obvious way

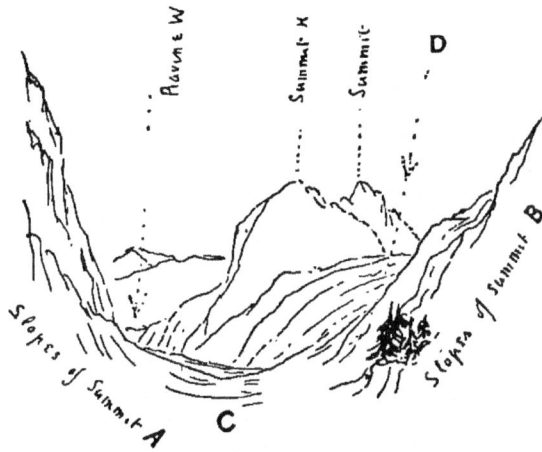

straight before him, and with nothing to tell him that he must go up a second col, two or three hundred feet above him to the right at D, if he is to get into the right valley.

It is in cases of this sort that Schrader's map is so useful—so far as it goes; but it only covers the quite central part of the Pyrenees, and the contours are 100 metres apart.

13

The particular ways in which one may get into the wrong valley are innumerable, but these three types which I have given include all the most common of them ; and, of the three, the last which I have described in such detail is at once the most perilous and the most common.

While I am upon this subject of getting into the wrong valley on the *downward* side, I ought to mention the tricks which the map and one's own judgment play upon one as one goes *upwards*.

Errors made as one follows the map *up a ravine* are nearly always due to making a false estimate of distance. The path may be lost for a considerable stretch, and the contours may at first be puzzling, but if one will trust to one's map and to one's compass one will never go far wrong, unless one misjudges distance, and it is on this account that in the directions I give below for particular places, I mean distance with what care I can.

Thus you may miss the path which branches off from the main path from the valley of the Cinqueta to go eastward over the Col de Gistain ; but if you have made an accurate estimate of distance, and trust to the measurements given, you cannot fail to identify the stream up which that crossing lies.

Nothing can replace judgment, but there is a rule of thumb which is workable enough, and that is, save under conditions of extreme fatigue, that your kilometre on a mule path hardly ever takes you less than twelve minutes or more than fifteen. I except steep

climbing of course, but steep climbing only comes at
the port itself, or in quite unmistakable ravines and
gorges, where you will not lose your way. Where
you lose your way is in the Jasse, or in the
bifurcation of main valleys, and there, as you plod
up your mule path, you will, as I say, never take
less than ten minutes over your kilometre (which is
a centimetre upon your map)—and you ought always
to have a little measure with you—nor will you
ever take much more than twelve, save when you
are quite knocked out and unable to calculate distance at all.

These limits will seem narrow to those who have
not experienced such paths. But they are wide
enough. You must of course note the times during
which you choose to stop, and it is also true that if
you make quite short halts for a moment or two, of
which you take no record, you will quite put out your
calculation ; but twelve minutes to the kilometre is
3 miles an hour, fifteen is 2½ miles an hour, and if a
man gets over a level mule track in the early morning carrying weight a little faster than the first pace,
or on a steep part at evening a little slower than the
second, yet the occasions when this rule of thumb
fails are rare.

When your watch tells you that by the distance measured you should be approaching a
bifurcation, or any other doubtful place, halt and
decide.

If you do miss your way going upwards, or do
take the wrong valley, if, in a word, you are lost (as
I was badly four years ago, so that I have the right

to speak of it), the first thing to remember is that the path, if you will take it *downhill*, will lead you at last to men. The rule about following running water is all very well in many mountains of the ranges, but it won't do in the Pyrenees, for the running water very often goes under sharp limestone cliffs, and if you don't find your way round or over them, you may spend more hours than are safe in looking for a way out. They form a very complete prison door, indeed, do these gorges.

The path, I say, if you follow it downhill, will save you, but if, when you find you are in the wrong valley, you attempt to recover your track by going up the lateral ridge, you always run a grave risk. It is by experiments of that sort that men die from exhaustion. It is true that one is not usually tempted to this extra effort. It is much easier to go on the way one is going, and to follow the path down, though one knows it is a wrong one, but there are occasions, especially late in the day, when one has *all but* conquered the main crest of the range, after perhaps one failure, and when one knows that one is lost, when the idea of one vigorous effort to get over while it is yet daylight is tempting. It is a fatal temptation.

When you have made up your mind that you are lost, or even when the map has told you so, pay no attention to anything else about you or within you, such as the guess that such and such a rock in front of one may hide such and such a village, or the hope that your strength will hold out for 12 or 15 hours without food, but at once behave like a person in grave danger, that is, calculate your chances of retreat, and think of that only, for I repeat, it is more easy to die from exhaustion than in any other way in these hills, and nearly all the people that perish in mountains perish from that cause.

When you have made up your mind that it is your business to find men again, and that you do not know how far men may be, first note your bread and wine and the rest, if any provision is left ; next determine to reserve it until nightfall : eat it then, do not blunder on through the darkness (it is astonishing what very little distance one makes after sunset, and every half hour of twilight makes it more difficult to camp) —sleep, and take the first half of the next day without food ; you are reserving your very last rations until the noon of that day.   For one can do a considerable distance without food if the effort is made in the early morning.

Never bathe under such conditions of fatigue, and towards the end, when you are exhausted, drink as sparingly as possible.

It is perhaps useless to give any hints about what a man should do when he is lost, because men get lost in mountains by hoping against hope and pushing on when common sense tells them to return. But I write down these hints for what they are worth.   After my first bad lesson in the matter, I found them fairly useful.   Remember, by the way, if you are lost and if there is no path apparent, that a cabane even in ruins somewhere in the landscape means a track visible or invisible, and that any rude crossing of the stream with stepping-stones or a log means the same thing. But you must not imagine that the presence or traces of animals will prove a guide, for even mules wander wild for miles on these mountains in places where a man can only go with difficulty and along random tracks leading nowhere.

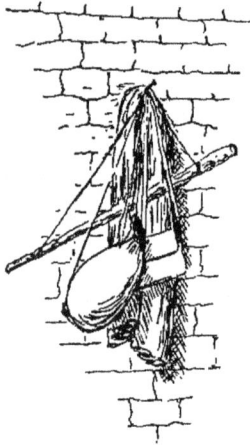

VI

THE SEPARATE DISTRICTS OF THE PYRENEES

FOR the purposes of travel upon foot, the range of the Pyrenees falls into certain divisions, which are not very clearly marked, but which arrange themselves in a rough manner under the experience of travel. As I come to deal with each of these, it will be seen that there is not one which does not overlap its neighbour, and it will be impossible to describe any mountain district without admitting this overlapping to some extent, because any valley connected by certain local ties with the valleys to the east and west is also, as a rule, connected with the valleys to the north or south of it. Still, the districts I speak of are fairly distinct, and consist in (1) the Basque valleys, (2) the Vals d'Aspe and d'Ossau, with the valleys of the Aragon and Gallego to their south, which I will call "the Four Valleys," (3) the Sobrarbe, (4) the three valleys attaching to Tarbes, to which I also attach the Luchon valley, (5) the Catalan valleys and Andorra—in which I include the Val d'Aran, (6) the Cerdagne (omitting the Tet and Ariège valleys), (7) the Ariège and Tet valleys, (8) the Canigou.

These I will take in their order, and I will begin with—

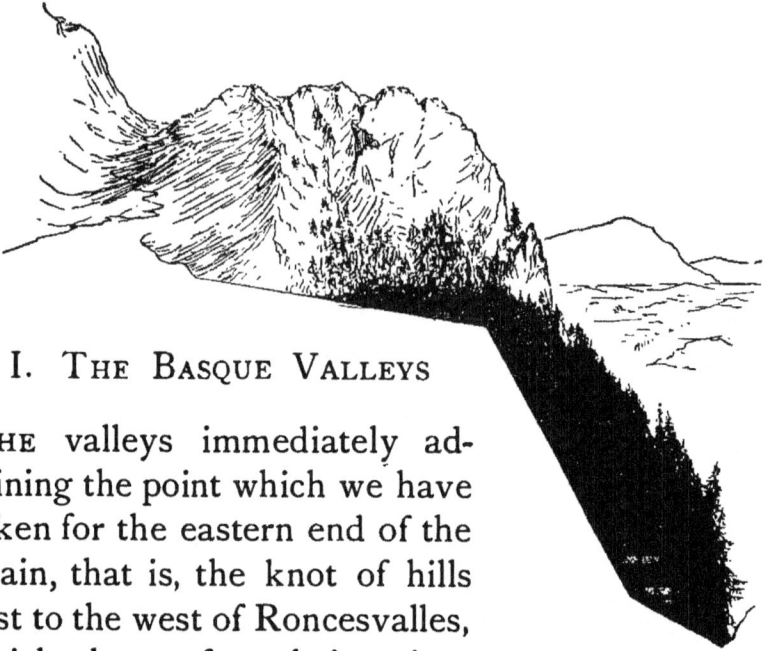

I. The Basque Valleys

THE valleys immediately ad-
joining the point which we have
taken for the eastern end of the
chain, that is, the knot of hills
just to the west of Roncesvalles,
which have for their pivot
Mount Urtioga, form one country-side and should
be considered together.

They are the Baztan to the west, the first of the
many valleys into which the main range splits up
like a fan as it approaches the Atlantic; the valley
of Baigorry, parallel to it and immediately to the east;
the valley called that of the St Jean in its lower
French part, and that of Val Carlos in its upper
Spanish one; this valley stands eastward of Baigorry,
and unites with it before leaving the hills to join the
valley of the Nive. The two together, and the
lower valley of the Nive, are called by the common
name of "The Labourd"; on the south of the range
comes the valley of the Arga and the plain south
of Roncesvalles: these make one division of the
Basque district. The same dialect of Basque is
spoken throughout the Labourd (there are variations
upon the Spanish side), the same type of house and

MAULEON

Aramits Val de Baretö Lourdii

Tardets Port Licq

aison

LES ARBAILLES

Alcayo Larrauo

SO

Pt. d'Ur-la R. Becal

Pic d'Orhy 6618 chagaria

pied de Port Lecumberry

R D FORÊT

Irissarr

t Étienne Blanca Borry

Val de Baz Mt Ah 36 urguete R. Iroti R. Urrobi rga Larrasoaña

4xe R.

of food and of hill is everywhere around. The other division of the Basque valleys is the French district of the *Soule*, just to the east with its corresponding valleys south of the frontier.

As to the Labourd and its accompanying Spanish valleys, the space open for camping or wandering in this corner of the chain is less than in the higher central part. The low round hills are often cultivated to their summits, the valleys are always well populated, roads and villages are many, and though there are one or two fine stretches of forest in which a man can spend as many days as he chooses (notably the forest of Hayra, which lies up southward at the far end of the Baigorry), they are not to be compared in extent or in wildness with the forests further east. The whole width of the Hayra, counting both the French and the Spanish slopes, is, at its greatest extent, not more than three miles. Its length is not six. The small lakes also that are characteristic of the Pyrenees throughout their length, are lacking here, and the prosperity and industry of the Basques press upon the traveller wherever he goes.

If one would stay some three or four days in this district, it is a good plan to leave the train at St Etienne, just at the beginning of the Baigorry valley. St Etienne is the terminus of the branch line which strikes off a few miles down the river from the line connecting St Jean Pied-de-Port with Bayonne, and one gets to St Etienne by the morning train from Bayonne about mid-day.

Immediately to the west of St Etienne, connecting

it with the Baztan, lies the pass of Ispeguy. It is of course very low, as are all these hills; it is little more than 1000 feet above St Etienne, or perhaps 1500, but from the summit there is a fine view of the higher distant Pyrenees to the east. The frontier runs here north and south, passes through the summit of the col, down the further side of which an easy valley road leads down on to the main high way of the Baztan.

This highway is the modern representative of the track which for many centuries connected Bayonne with Pamplona. It was, until recent times, a mountain way; the main Roman road went through Roncesvalles. It is now, as was seen when we spoke of roads for driving and motoring, the best approach from the French Atlantic coast into Navarre. From the point where you strike this high road, where the valley debouches upon it, and where the lateral stream you have been following falls into the river Baztan, there is a walk down to the left, or southward, of some 4 miles, into the town of Elizondo, which means in Basque " The Church in the Valley." For the Basques, like the Welsh, have the terms of their religion mainly in the form of borrowed words, and the Greek Ecclesia, which is " Egglws " in the Welsh mountains, has nearly the same sound here, 800 miles to the south, and with all those days of sea between. Christendom is one country.

There is no easy journey from Elizondo down to the south of the hills and back east again into the French valleys, unless you go on to Pamplona, although of course there is nothing high or steep

to stop you, if you have plenty of provisions, except
the absence of maps (which do not exist for this
district upon any useful scale to my knowledge).
If you want to make a mountain journey of it without touching the town of Pamplona, go down a mile
or two from Elizondo to Iruita, where the main road
branches into two ; thence going south and a little
east up the stream which comes down from the
frontier summits, you may go over a col between
that valley and the valley of the Esteribar, where the
Arga rises. You will find yourself at the first little
Basque village, that of Eugui, by evening ; the total
distance from Elizondo to Eugui, if you go the
shortest way, is only 20 miles. But, I repeat, it is
a difficult job. Maps are lacking, the valleys have
many ramifications, and the first part of your journey
is all up hill for half the day. If the weather is
cloudy it is more than possible that you will get into
the wrong valley, and find at last, when you have
got over your col, and are following the running
water on the further side, that that running water
is not the Arga at all, but one of the streams that
lead you back again into the Baigorry. However, if
you make Eugui in the Estribar, the rest is simple :
there are villages all round, connected by paths, and
not more than a mile or two from one another, and
you may go through Linzoian to Espimal and so to
Burguete, where you get the main road over Roncesvalles, without fear of losing your way ; for there
are people everywhere.

It is best, however, when you have slept in Elizondo,
which is a very pleasant little town, to take the

motor bus and get on to Pamplona ; for the Basques, who detest as much as the Scotch to be behind the world, have a motor bus along this mountain road. From Pamplona next day you can go by the new road to Burguete, passing through Larrasoaña and Erro. It is a long journey of nearly 30 miles ; it can be broken, if you choose, at Erro, but the sleeping accommodation there is nothing very grand. If you push on beyond Burguete, over Roncesvalles, you can, in something under 40 miles, get to Val Carlos, the last town in Spain, and for those who can walk 40 miles this is the best thing to do. If not, break the journey in two at Erro, desolate as the little place is.

The object of course of this walk is the Pass of Roncesvalles, and the vast contrast between the slightly sloping Spanish plain of Burguete, running up to the summit of the Pyrenees, and the great chasm which opens beneath your feet when you have reached that summit, and which forms the entry into France.

You will not easily make a camp in any part of this round, and it is well to remember here, where first mention is made of crossing the Spanish frontier, that the Spaniards will not let a man leave their country unless he has due permission upon a paper form. Why this should be so I do not know, and I have very often gone in and out of Spain without telling the authorities, as I have for that matter gone in and out of Germany on foot, though the German officials are more stupid than the Spaniards, and therefore attach much more importance to such things.

Still, it is safer to ask for your permit, and it will be given you by a functionary called a " Corregidor," at Val Carlos. A few miles beyond, eight to be exact, you are in St Jean Pied-de-Port, which is the head of the railway to-day, and which has been for nearly 1000 years, the depot town at the foot of the pass for armies and for travellers. On this same flat where it stands, was the Roman fort and depot, but not quite on the same place ; it stood on the spot now called St Jean le Vieux, 2½ miles up the lateral valley. This last was the halting place of Charlemagne in the famous story, and St Jean, as we see it, is a town not of the Dark but of the Middle Ages.

The next district to this of the Labourd, lying immediately to the east of it, we have seen to be called the Soule. It is also Basque, though it is Basque spoken with a different accent, and with certain verbal differences as well. The way from one to the other lies through wilder and more likely land for camping than is to be found in Baztan, Baigorry, or Roncesvalles. It is a good plan, if one has the leisure, to approach the Soule on foot by way of St Jean, though the more ordinary way is to go round through the plains by train to Mauléon (which is the capital of the Soule.)

If one goes on foot directly across from the Labourd into the Soule, he strikes that valley in its higher reaches, and well above Mauléon.

The shortest line, if one does not mind sleeping in a mountain village, is to take the high road from St Jean Pied-de-Port to Lecumberry, and to follow that

way up the valley of the Laurhibar until the high road
comes to an end.   It did so abruptly two miles or so
beyond Laurhibar, some years ago, but as it is being
continued, one may follow it every year further up
the dale.   The high road ends (or ended) about 10
miles from St Jean; and Lecumberry is the last *village*
still, however far the road may have progressed up the
valley.   When the road ceases one must continue up
the valley by a path on the left bank of the stream.
One soon finds on this left bank a series of precipitous cliffs ; one must there cross over to the path
upon the right bank.   It is also possible to keep to
the right bank all the way—there is a track on
either side—but I speak of the usual way.   Hence-
forward the path remains quite clear and runs close
alongside the stream, with steep cliffs upon the further
shore, until, in the last mile or two, before the head
of the valley, one enters a wood, and it is here that,
if you are not very careful, you will lose your way.
The contours are complicated, the valleys numerous,
and the alternation of wood and open land most
confusing.   But if you will go *due east* by your
compass from the point where you entered the wood
(abandoning the path where it crosses the stream
and goes over to the south), and if you will remember always to turn any precipice or ledge of rock by
descending to the *left* of it, and always to *descend*
after you have made the first high open space, you
will come upon a clear track not quite three miles
from the point where the path enters the wood.

It sounds but a vague indication, but it is a
sufficient one, because bad precipices prevent you

from going too much to the right, and the natural tendency of man to go down hill when he can will prevent you from going up on to the ledge upon your left. You will find yourself shepherded—if you always go as due east as is possible, and always turn a ledge of rock to the *left*—into a track which runs all along the high lands above the slopes that dominate the Brook Aphours; a little way down, that track falls into a high road, and a few miles further the road reaches Tardets, the central town in the valley of the Soule, half way between Mauléon and the highest summits. The whole journey from St Jean thus described is a big distance, nearer 40 miles than 30, with windings all the way, and you must be prepared if you become fatigued or have bad luck with your weather, either to camp out in the woods at the summit of the pass, or to sleep in the first hamlet upon the eastern side.

There is, indeed, a short cut which strikes the valley much higher, but it is difficult to make and involves the climbing of two cols. For this short cut the directions are as for the last, until your path along the Laurhibar has struck the wood; there, instead of leaving it when it turns south, and instead of going east (as above), you must keep to the track. It will cross the stream, still going due south, wind up between an open space through the woods, and will point before you lose it to the climb over the shoulder of the Pic d'Escoliers; it is a stiff climb of nearly 2000 feet from the point where you crossed the stream and very steep. The 2000 feet or so are climbed in under two miles.

When you get to the shoulder of the peak a steep
southern slope lies before you, diversified and made
perilous by rocks, and separating plainly into an
eastern and a western valley. Between you and the
eastern valley (which is that you must descend) are
steep rocks ; they can be turned, however, by going
to the *right* of them, but the whole place is precipitous and difficult. The advantage appears when
once you are down on the floor of the valley (which
is but 1000 feet from the peak), for you come within
a mile to a clear path, and once you have come to
this, you are, in another two miles, at the village of
Larrau, which is the terminus of the great national
road, and stands in the last upper waters of the
valley.

If you approach the Soule by the more ordinary
way you will come by train through Puyoo, change
there, and take the train for Maulèon ; and Maulèon,
as I have said, is the capital of the Soule. But the
true mountain town is Tardets, half way up the
valley. Tardets is the market town for all the
Basques of the hills, and you can never have enough
of it, both for its heavenly hotel, of which I shall
speak when I come to speak of hotels, and for its
universal shops, and for its kindly people. It stands
in an opening of the lower hills, just before the
valley narrows and enters the high mountains, and
you may reach it from Maulèon by a tramway which
runs up the river as far as Tardets and then turns
off to the left and goes round to Oloron.

If you approach the Soule in this manner, making
Tardets your starting-point, you will do well to

equip yourself in that town and then to continue up'
the valley some five miles past Licq, until you come
to the fork of the river. It is an unmistakable
point, because a very definite rocky ridge comes
down and separates the two sources of the river
Saison, which is the river of the Soule. The
branch to the right (as you go southward) leads up
the valley to Larrau, of which I have just spoken,
and the high road follows it ; the one to the left
(which is the main stream and is called the Chaitza)
has no main road along it, but a good mule
track, very clear and plain, and leading at last to
the village of Ste Engrace, which lies at the
extreme end of the valley and gives the whole
district its name.

Ste Engrace was a saint of the persecution of
Diocletian. She was martyred in Saragossa, and
the name of the village is one of the many examples
of the way in which the southern influence overlaps
these hills. I have said that the Spanish sandal is
used to the very foot of the French Pyrenees, and
so is the wine-skin which is common to all Spain,
and so is the Spanish mule. Here you may see
the Spanish saints as well reaching beyond the
summits.

From where you leave the main road and go up
the Chaitza valley to Ste Engrace is a distance of
8 or 9 miles, and in this valley, in its upper waters,
is to be found one of the wonders of the Pyrenees,
and also one of the main passages into Spain.

The wonder is the gorge of the Cacouette ; the
passage is the twin passage of the Port d'Ourdayte

and the Port Ste Engrace, and near them to the west are two easier ports.

The Cacouette is a cut through the limestone such as you might make with a knife into clay or cheese, with immense steep precipices on either side, and apart from the track above the cliffs there is some sort of tourist's way along the cavernous ravine for those who admire such things. Of the two ports, the one path goes up the western side of that cleft in the limestone (which drops 1500 feet into the earth), and the other goes up the eastern side To take the road up the western side, you leave the Ste Engrace road 3 miles after leaving the great highway, by a lane which goes off to the right and drops down into the valley ; it is quite plain, and is the only road so leaving the main track, so that it cannot be mistaken. It climbs the opposing hill, and if you follow it through all its windings it will take you to the Port Belhay, or to the Port Bambilette, both under a mountain called Otxogorrigagne, and both easy. But if you continue just above the limestone precipice, you will come into a very striking circus of rock just under the watershed, up which your path perilously climbs to the summit and the frontier ; this is the Port d'Ourdayte.

The Port Ste Engrace, though not half a mile distant from it, is reached in quite a different manner, and the separation between the two is due to this limestone gorge, which cuts off one path from the other.

If you are going to try to cross by Ste Engrace,

sleep at the village before starting. There is a good comfortable inn kept by people of the same name as those who keep the inn at Elizondo, Jarégui. It is so steep and difficult a bit that if you were to attempt to do it in one day, without sleeping at Ste Engrace, you would hardly succeed unless you already knew the mountain well, and mist, which is the fatal difficulty of these western Pyrenees, will more commonly catch you in the early afternoon than at any other time in the day, so that you had better make your ascent before noon. When you have slept at Ste Engrace you will find the path the next morning winding round through the woods, at the base of the hill opposite the village. One must ask the way to the start of this path, and it is not always clear after the first two miles; one has now and then to cast about for it a little, but at last it emerges upon a high grassy slope, which runs all the way to the crest of the hill and the frontier. The path does not follow the straight ascent of the hill, it curves nearer and nearer to a precipice which is the same as that climbed by the neighbouring paths of the Port d'Ourdayte; for ten dangerous yards it runs on a tiny platform right along the gulf and makes over the crest into the further Spanish Basque valley, whose capital is Isaba.

Of this valley I can say nothing, for I have not succeeded in crossing the Ste Engrace, though I have twice tried, but I am told that Isaba is among the best of these little mountain Basque villages or towns for entertainment and for cleanliness, and all Basque villages and towns are cleanly. There is a good posada. From Isaba also a high road runs into the higher valleys of Navarre and to Pamplona.

Near this territory of the Soule, and partly included in it, are two great districts where a man may spend many days at his ease in camp there. The first is the great forest of the Tigra, which stretches to the east of Tardets and is full of rocks, rivers, and adventure. You may take it at its greatest width, counting one or two open spaces, to be 8 or 9 miles, and at its greatest length, from the Peak of the Vultures to St Just, to be much the same. Its high places, some of which are bare peaks, some clothed with woods, range for the most part round about 3000 feet, but the highest point—of which I have never heard the name, and which is on the very south of the forest, just passes 4000 feet. Tardets is always at hand on the one hand, St Jean Pied-de-Port rather further on the other ; from both one may re-provision oneself.

Another and still larger district lies on the further side of the valley to the north and east of Ste Engrace itself. It is the great mass of wood, mainly beech, which stretches all over the hills between this last Basque valley and the Val d'Aspe, next to the west, which is the frontier valley of Bèarn. These woods have no common name, they are intersected by clear spaces, notably round the higher peaks of the forest, but they make a district of their own stretching eastward and westward from Lourdios to Licq, northward and southward from the frontier nearly to Lanne, and thus measuring not much less than 10 miles every way, in French territory alone.

There is no forest in which it is easier to lose one's way than this great stretch of upland. This is especially true in the Suscousse district, due east of Ste Engrace; there is here a labyrinth of complicated valleys, and what seems on the map so easy a passage from the Soule into the Val d'Aspe is in practice nearly impossible to find. To camp in and to explore, this forest is even better than the Tigra; for its summits are higher, and its views more unexpected and remarkable. There are points in it which are more than 6000 feet in height, and the great Pic d'Anie, the first of the really high mountains of the chain, stands high above them, just beyond the southern limit of the trees.

II. The Four Valleys (Béarn and Aragon)

FOUR valleys in the Pryenees count together in travel upon foot. They are the Val d'Aspe and the Val d'Ossau on the French side, and the valleys of the rivers Aragon and Gallego on the Spanish side.

These four form a unity for the reason that in one place (which is just to the south of the watershed) they are, without too much difficulty, approachable one from another.

Many historical accidents have also served to unite these four valleys. One pair of them made the platform for that great Roman road to which allusion has so often been made in this book, and which ran from the French plains over what is now called the pass of the Somport, right down through Jaca to Saragossa. The parallel pair of valleys just to the east, the Val d'Ossau and the valley of the Gallego, on the Spanish side, though no high-

Na

N

Rébénac

Val de Baretou

rudy

Lauzon A

Asasp,

urbe

Bielle o

Lourdios

s
s
o
O
d'

Laruns

FORÊT d'ISSAUX

Bedous

Gave d

ccous

aux Bonnes

Les Eaux Chaudes

Va le

ave

Gab

Urdos

t d'Anso

✳ C. de Pao

Fuerte

Troundael
5768

✳ Col du
Roya   canal Roya

Sallen

Passde
5499

V del Canfranc

Saldinies a

Va del Hecho

Canfranc

e Tena

Hecho

R. Aragon Subor

R. Esterro

sa

Val d'Aragon

Biéscas

way ran along them until quite recently, had a similar historical unity which bound them both together, and bound a pair of them to the two sister valleys upon the west. For the eastern part of what later became the kingdom of Aragon, the county of Sobrarbe, stretched from the valley of the Gallego eastward, and was a natural line of defence southward against the Mahomedans; while the Val d'Ossau to the north of it was reached by an easy pass and must have formed—though we have no exact historical record of it—a good road for the parallel advance of armies.

It must never be forgotten that when an army is advancing in great numbers it is of paramount importance for it that the host should be able to concentrate before action. But roads, especially roads over mountains, compel men to march in long strings, so that the head of the column will have arrived at a particular point hours before the tail of it; and what is more, the deployment of the column, that is the getting of it all into a front perpendicular to its line of advance, takes time in proportion to the length which the column had before it began to deploy. This accident it was, for instance, which destroyed the French and their allies at Crecy, for though they greatly outnumbered the English they had come up in columns too long to deploy in time. Now it evidently follows from this principle that armies on the march, even under the rudest conditions, will attempt to follow parallel roads. To find two roads parallel to one another and leading to the same field of action is to halve the difficulties

of transport and of deployment.   But it is very difficult
(under primitive conditions) to find two parallel
roads which are near to one another, and unless the
lines by which the army advances are near to one
another the advantage of the alternative routes will
disappear in proportion to their distance one from
the other.   In mountain regions it is especially
difficult to find two passages parallel to each other
and yet in close neighbourhood.   This is precisely
the advantage afforded by the trench of the Gallego
continued in the Val d'Ossau to the east, and in
the trench of the Aragon continued in the Val
d'Aspe to the west.   Two hosts using the old mule
paths could leave Sallent on the Gallego and Canfranc
on the Aragon at dawn of one day, and both would
meet at Oloron in the French plains before the
evening of the morrow ; on the southward march a
host could assemble in the plains of Bearn, separate
to use these two easy passes, and meet at Jaca at
the end of the second day.

It is fairly certain therefore—much more certain
than a thousand of the historical guesses that are
put down as truths in our text books—that the easy
pass between the Gallego Valley and the Val
d'Ossau was twin throughout the dark ages to the
great Somport pass not 8 miles westward of it.
Abd-ur-Rahman must have used both and so must
the Christian knights when they came so often to
the relief of Aragon in the heavy and successful
fighting against Islam which marked the tenth and
eleventh centuries.

To appreciate how close these two parallel tracks

were to each other one has but to remember that
the gap between the Val d'Aspe and the next easy
pass westward—right away at Roncesvalles—there
is a matter of 40 miles.   Between the Val d'Ossau
and the next easy pass eastward there is a gap of
indeterminate length according to the definition of
the term "easy," but there is at any rate no notch
over which one could take any armed force until one
gets to the Bonaigo, quite 60 miles away.   All
between is the mass of the highest and most rugged
ridges of the Pyrenees, over which certain paths
have always existed, indeed, and over which, in two
places at least, at Gavarnie and at Macadou, the
French propose to drive roads, but no gap in which
was ever passable in the Dark and Middle Ages for
a great number of men.

I have said that these two parallel trenches were not
only twin in history for the use of armies, but were also
communicable one from the other just south of the
watershed.   North of it, indeed, the Val d'Aspe and
the Val d'Ossau, though one can be reached from
the othèr, only communicate by very high and rocky
ridges, the easiest of which is the Col des Moines.
But on the south side there is one accidental easy
passage.   You may go all the way from the Somport to Jaca and find nothing but the most difficult
mountains on your left, and all the way from the
Tourmalet (which is the pass at the top of the Val
d'Ossau corresponding to the Somport) down to
Sandinies and find nothing but difficult mountains
on the right, save just at the beginning of the descent
where this accident of which I speak occurs.   Its

feature is a lateral valley called the Canal Roya which takes its name from the streak of intense red scoring the side of its principal peak.

This lateral valley points right away eastward from the trench of the Aragon, it is nowhere precipitous along its stream (a rare advantage in the Pyrenees) save in one spot where a quite low precipice is easily out-flanked along the grassy slopes above it. And the end of that valley consists in a sort of semi-circular ridge of grassy steep banks in three places of which ridge, at least, a man or a beast can walk over without difficulty or danger. These three places are the Port de Peyreguet, the Port d'Anéou, and the col of the Canal Roya. This last is the principal one, the easiest and the lowest. Each is within half a mile of its neighbour, and on the further side one comes down quite easily by large steep slopes of meadow to the valley of the Gallego. The Port de Peyreguet and the Port d'Anéou bring one down just on the north of the flat dip of the pass, the col of the Canal Roya just on the south of it ; but whether one comes down just north or south of the flat Pourtalet pass is an indifferent matter. The travelling in all three cases is little more than a walk.

These "gates" up the Canal Roya from the Val d'Aragon into the parallel valley of the Gallego knit the whole four valleys into one system, and to this day their customs and their inhabitants have very much in common, and the two valleys, which were the core and heart of Aragon and the origins of its crusade southward against the

Mahomedans, count in history and in local geography with the two valleys which were the heart and origin of Bearn up to the north.

The Val d'Aspe, which is the most important of the four, is that valley in the Pyrenees where the characteristics of the range are most strongly marked. It might serve as the type of all the others. You cannot see the opening of it southward from Oloron without appreciating that you are approaching something distinctive and singular in landscape. It is so clean-cut and so obviously an invitation to the crossing of the hills. The gorges which divide into separate flat steps every Pyrenean valley, are nowhere more marked than here. The village of Asasp which stands at the first of them is singularly characteristic of such an entry; the gap through which the old lake broke is so clear, the walls through which the Gave runs are so perfect.

Somewhat further on when yet another gorge has been passed there opens out one of those circular and isolated spaces of which Andorra is the historical example, and which in greater or less perfection are characteristic of all these hills.

This plain, which still recalls in its contours the old Lake which created it, and of which it is the floor, is more regular and more complete than any of the many *jasses* and "*plans*" which distinguish the other vales. It is even more striking than that of Andorra. It nourishes five villages which might easily (had not the great international road run through them for 2000 years) have federated to form an independent Commonwealth as the eight villages of Andorra federated to form one. Indeed this circus, surrounded by almost impassable hills which meet at either end in narrow Thermopylæ, was very nearly independent at the close of the Middle Ages, and when it appealed against the king for the preservation of its customs, these were preserved by the authority of the king's court.

Of the small towns or large villages which this little secluded corner of the world contains, Bèdous is that which will seem the capital to the wayfarer for it is the only one which stands upon the main road; it is the terminus of a railway which will soon be finished, and of which nearly all the track is already made Bèdous, by this time, must also have more population, as it certainly has more wealth than any of the surrounding places. But Accous is the true capital of the five, and it is pleasurable to hear with what reverence the villagers of the farms around speak of Accous as though it were an Andorra-viella or a Toulouse. All this wonderful and silent plain is marked with long lines of poplars which enhance by their straight lines the immensity of the heights around them.

If one will pass some days in this singular valley it forms an excellent place from which to explore the high passes into the Val d'Ossau, and the bases of the two great mountains which, to the east and to the west, neither visible from the floor of the valley, are, as it were, its guardians : the Pic d'Anie and the Pic du Midi d'Ossau. The man who does

THE PIC DU MIDI D'OSSAU

not desire to cover much ground but who wants thoroughly to know some very Pyrenean part of the Pyrenees will do well to stop at the Hotel de la Poste at Bèdous, and thence climb at his leisure up on to the platforms from which spring these isolated and dominant masses of rock.

The Pic du Midi remains in one's mind more perhaps than any of the isolated mountains of Europe. It is quite savage and alone, and you must fatigue yourself to reach it. There is no common knowledge of it and yet it is as much itself as is the Matterhorn. The Pic d'Anie, though it is less isolated, stands even more alone and has this quality that it dominates the whole of the seaward side of the Pyrenees for it is much higher than anything westward of it. Also it is the boundary beyond which the Basques and their language have not gone.

Beyond this plain of Bèdous, when you have passed the southern "gate" of it, you come into a long, deep and winding gorge which leads you at last to Urdos, and Urdos is and has been since history began the out-post of the French in these hills. It was the Roman out-post and the medieval one, and it was the out-post through the Revolutionary wars.

Napoleon, who in everything recognised and imitated the example of Rome, and who, for that matter, caused the Empire to rise again from the dead, determined that a modern road should go again where the old Roman road had gone. He determined this in connection with his Spanish wars, and decreed in 1808 that a way for artillery should cross where the legions had gone. But Europe, as we all know, would not upon any matter accept in the rush of a few years the constructive desire of Napoleon and of the Revolution. It has taken more than three generations to do not half the vast work they planned, and this road, which like almost every good road over the Alps and the Pyrenees has Napoleon and the Revolution for its origin, waited till past the middle of the nineteenth century before it reached so much as the summit of the port.

Under Napoleon III., in the sixties if I remember right, the thing was done and the road reached the summit of the Somport, the lowest and the most practicable of the high passes of the central mountains. But the Spaniards still hung back and it was not till the other year that the road upon the Spanish side was completed. Now, however, one may not only go all the way upon a high carriage road from Oloron to Saragossa straight south across the hills, but one may find the whole way marked with mile-stones as the Romans would have marked it, and saved at every difficulty by engineering of which the Romans themselves would have been proud. Once over the summit there is no resting place till one reaches Canfranc, 6 or 7 miles by the windings of the road below one. After Canfranc the valley of the Aragon, which one has been following, opens, and the plain of Jaca lies before one bounded by its great ridge to the south-ward, the Peña de Oroel.

If one would not go all that length of high-road (from Oloron to Jaca is over 50 miles) there are upon the Spanish side two lateral diversions which a man may take. The first is over the Col des Moines, the other into and over the Canal Roya.

The first can be seen right before one at the summit of the pass ; for when one stands upon that summit one has, running eastward from the road, a great open valley at the head of which is clearly distinguishable a bare rocky ridge with a low saddle which is the Col des Moines. It is perfectly easy upon either side, and upon the further side it shows one the splendid and unexpected vision of the Pic du Midi standing up alone beyond the little tarns at its feet : a double pyramid of steep rock upon which the snow can hardly lie in tiny patches and whose main precipices are dark, to the north, away from the sun.

The next lateral valley southward of the Col des Moines is that of the Canal Roya, but one can only enter it after going down the main road for quite a thousand feet. There a bridge will be seen spanning the Aragon and a little doubtful path leading beyond eastward up the lateral valley. It is two hours up that valley to its head by a path going first on the right bank of the stream then crossing over to the left one. One thus reaches by a continuous ascent the cirque or amphitheatre which bounds it at the eastern extremity of the valley. Here there is a difficulty in finding the easiest and lowest col. The map is doubtful and the details upon the map are not sufficiently

numerous.   The Canal Roya is well worth camping
in and returning by to the main Spanish road if one
is inclined (and if one is, one would do well to camp
near the wood upon the left bank of the stream not
quite half way up the vale for there is no timber
further on).   But if one does not camp and prefers
to get over the col into the valley of the Gallego
the rule is to note a sharp peak which stands exactly
at the apex of the valley—it is the lowest of the
peaks around but very distinct, forming an isolated
steeple due east of the last springs of the stream.
The way lies to the left or north of this peak and
just under its shoulder up a loose mass of fallen
rocks on which an eye practised in these things can
discover from time to time a trace not of a true path
but at least of infrequent travel.   Upon the far side
easy slopes of grass take one down in about an
hour to the Sallent road.

Note that these two cols and the stretch from
road to road and from inn to inn can only with
some peril be undertaken in one day from Urdos.
In fine weather and without accident the thing is
simple enough, but when you are baulked for an
hour or two by the trail, or if you start a little late,
or if you are detained by mist you may very easily
not manage the passage from one of the great roads
to the other, near as they look upon the map.

With everything going well carrying little weight
and fresh, it is quite three hours (and more like
three and a half) from Urdos to the bridge over the
Aragon.   It will be another two up the Canal Roya
and two more over its col and down the other side

to the high road, and even from that point on the high road, if you follow the road only, there are two more hours before you reach Sallent. It is a very heavy day of quite 30 miles with two cols, one of 5000 feet the other of 6500 feet, to be taken on the way, and it is foolish to undertake either the Col des Moines or the Canal Roya from Urdos without allowing for the chance of one night at least upon the mountain.

The second pair of valleys, that of the Gallego on the Spanish side, and the Gave d'Ossau on the French side, are linked together by two very easy passes, and one difficult one of which I shall speak in a moment.

The old port, now called " Port Vieux de Sallent," or the " Puerta Vieja," is easy enough, though it went over a higher part of the mountain than the new pass just next door to it. I say it is *higher* than the pass now used, and this contrast is not infrequently found in the Pyrenees, some feature or other in the topography of the ridge making it more convenient for a native to cross by a slightly higher saddle than by some lower one close by. For instance, the Somport itself is somewhat higher than a quite unknown gap four miles to the west of it, but this lower gap was never used because it led into a Spanish valley of a difficult and most isolated kind.

In the case of the two passes from the Val d'Ossau into Spain, the obstacle which prevented the lower pass being used until quite lately, was a great mass of rock overhanging the sources of the Gave d'Ossau, in the highest part of the valley. When the new highway was made, this rock was blasted and cut so as to take the road round it, and thus the low pass beyond, called Pourtalet, was utilised. It is below 6000 feet and exactly 1000 feet lower than the old Port de Sallent. But even nowadays, if you are on foot you will do well to cross by the old port, high as it is, for it saves time.

While I am on the subject I must warn the reader that the $\frac{1}{100,000}$ map does not accurately convey the shape of the last two miles of the road upon the French side, and the line of road mere guesswork upon the Spanish, though the shape of the mountains is accurately given.

This pair of valleys is remarkable for another feature upon the French and upon the Spanish slopes : their wildness. Let me speak first of the French. The French valley, the Val d'Ossau, is one of the wildest and most deserted in the Pyrenees, and also it is the one most densely clothed with forests. The reason of this is that there is less flat ground at the foot of it than in any other. Nowhere does it expand into even a narrow circus, and about Laruns, where it debouches upon the lowlands, and the summit of the pass into Spain, a distance of perhaps 17 miles, there is but one large village, close to the bottom of the valley, and that owes it existence to Thermal Springs ; it is called Eaux Chaudes—a dismal place, squeezed in between the torrent and the cliff, dirty, uncomfortable, and sad. Higher up, however, a tiny hamlet, the humblest and most remote in the world, one would think, has of recent years taken on some little importance through travel; this is the hamlet of Gabas, which may be said to consist, in three inns, a ruinous chapel, most pathetic, and a customs station. Of the excellent inn at Gabas, I will speak elsewhere.

This valley of the Ossau is the base for two districts, both of which are very Pyrenean, and on either of which a man may spend a day or a month of lonely pleasure. One is the steep and very fine valley of the Sousquéou, the other is the short and extremely steep torrent bed which leads up to the foot of the Pic du Midi.

This mountain dominates all this section of the Pyrenees. The approach to it by the Col des Moines I have already mentioned; this ascent by the short valley from Gabas, through the woods, is better, because you come right up on to the mountain suddenly from the depth of a vast forest, and you feel its isolation.

I know of no hill which seems more to deserve a name or to possess a personality. Round its base there is matter for camping for days or for weeks, good water, lakes to fish in, shelter, both of rocks and of trees, human succour not too far off (Gabas is not three miles as the crow flies from the summit of the mountain), and a complete independence.

The Sousquéou is a less human excursion, though it has a very fine lake at the head of it. The communication with men is steeper and more difficult than from the district surrounding the Pic du Midi, and, as I know from experience, it is not difficult to lose one's way. Moreover, the exits from the upper end of this valley are not easy, and it is bounded on either side by the most savage cliffs in the whole chain. Should it be necessary to escape from this ravine by any path but that which leads down on to the high road near Gabas, you have no choice but the high and steep Col d'Arrius, which brings you down into the upper valley of the Gave d'Ossau, or on to the very high and most unpleasant Col de Sobe, which gets you into one of the most difficult parts of the Spanish side near the Peña Forata and so down to the Gallego. Its very remoteness, however, and its partial changes, may attract one kind of walker to the Sousquéou, but if he attempts it, let him go with at least three days' provisions. There are huts in the lower part of the valley, but there is no very good camping ground near the lake I believe, save on the side of the wood to the north. It is a lonely place, not without horrors, and is perhaps haunted; the shape of the hills around is very terrible.

The Spanish side of all this is more simply described, the new high road runs down 8 or 9 miles to Sallent, which can be turned into 5 or 6 miles by taking the old mule track that cuts off the windings of the graded road. The river Gallego runs below and increases as it goes. To the right or westward of the valley there is nothing in particular to be done, there is but one place where you can conveniently cross over into the valley of the Aragon, which is the Canal Roya I have already described; south of that crossing the flank of the mountain lies bare and open affording neither camping ground nor interest. On the left are the curious serrated precipices of the Peña Forata, where climbing makes but a day's amusement, but where also there is no opportunity for camping, and once Sallent is reached, though the "valley of Limpid Water" which runs north of it is fine enough, there is little to be done but to go on to Panticosa. There is a path over the very high ridge of the Pic d'Enfer, and there is a main carriage road which goes round the flanks of that mountain.

All this part the valley of the Gallego is bounded by some of the highest and most abrupt peaks in the chain, and (as I shall presently describe) another district, meriting another type of description and travel, lies to the eastward, and constitutes those new fortresses of the hills, the roots of old Sobrarbe, where Christendom first began to hold out against Islam, and whence the men of Aragon could securely push southward when the advance to the Reconquest began.

III. SOBRARBE

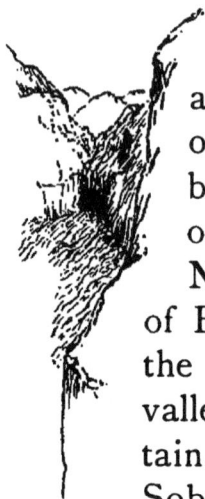

WHEN one says Sobrarbe one means all that eastern and larger part of the original valleys of Aragon which lie between (and do not include) the valley of the Gallego and the valley of the Noguera Rabagorzana, that is, the valley of Broto (which is that of the River Ara), the valley of the river Cinca and the valley of the river Esera; for, with certain ramifications, these three make up Sobrarbe.

That part of it of which I shall here speak, the part right up against the frontier ridge, is included between the big lump of mountains which surrounds Panticosa (of which the Vignemale is the most conspicuous) and the other big lump of peaks which is called the Maladetta group.

It has three towns corresponding to its three valleys, Torla in the Broto upon the Ara, Bielsa upon the Cinca, Venasque upon the Esera.

The Cinca, however, receives, right up at its sources, an affluent longer and more important than itself, called the Cinqueta, and on this stream is a group of villages, none of them important enough to be called a town, but standing so close together as to make a considerable centre of habitation.

THE SOBRARBE

Pic de Sauv 8051
Pt de Ve 8983
MALADETTA 11.168 & Nethou P.
R. Esera
Br. of CUBERAS
Venasque
Pt. d'Oo 9846
Val de Astos
Val de Louron
Aiguestoltes
P. des Posets 11.047
Val de Venasque
C. de Sahun
Sahun
St. Juan
Pt. du Plan 8061
C. de la Cruz
Gistain Plan
Pic d'Ourdissettou
Riomajou
Cinquetta
Neste d' Aure
Pass de Bielsa
Val d'Aure
Pic d'Ourdissettou
Barrosa R.
Bielsa
Salinas
Sarabillo
Cuesa R.
Plan de Bielsa
Pass de Barrouda
Cirque de
Pt. de la Munia
Toumous
Olos de Bielsa
Rivalta
Escuain
R. Vellos
Pragnouet
Heas
Jedre
Bareilles
Olascartez
Val de Pinede
C. d'Anisclo
Col de Escuain
Cinca R.
10334
Mt Perdu 10094
C. de razas
Val d'Nisc
Gavarnie
Cirque de Gavarnie 9200
Brèche de Roland
Pt Gavarnie
Bouchero
Pt du Vignemale 10820
BRIDGE OF N of NAVARR
Torla
Broto
Val de Broto
R. Ara
G. du Pau

Political Boundary
Heights in Feet 11.047

Railways
Tramways

Main Roads   Tracks ---
Other   do.

of Miles
0   5   10

But for these towns, the group of villages I have mentioned and one or two tiny hamlets, these Spanish valleys are wholly deserted, and they form by far the most rugged and difficult district of all the Pyrenees.

They also hold the highest peaks of the mountains; the culminating Nethou Peak of the Maladetta group, just upon the eastern edge of the district, (11,168 feet); the Posets (11,047) the Mont Perdu (10,994) the Pic d'Enfer (10,109), the Vignemale (10,820) all stand here. Most of the high peaks are in Spain, but it is another feature of the district that the frontier ridge is higher here than in any other part, and is also more continuous. The summit of the Vignemale forms part of it, and the notches by which it may be traversed in these 40 to 50 miles, lie but very little below the surrounding peaks, Only 3 of the passes miss the 8000 foot line. The Port de Venasque, at the extreme eastern end opposite the Maladetta, is 7930 feet in height; the Port de Gavarnie at the extreme western end is 7481. These two form the chief thoroughfares over this high and difficult bit; that of Gavarnie, upon the French side, is being prepared for wheeled traffic. The third, the Port de Pinède, also misses the 8000 foot line, but only misses it by 25 feet. All the other passes are but slight depressions in this barrier of cliff. The Tillon or rather the Passage to the side of it, is little under 10,000 feet, the Pla Laube is over 8000, so is the Marcadou, so is the better known and more used pass of Bielsa, while the Port d'Oo, is 9846 and the Portillon d'O is 9987.

The impression conveyed by this long line, the only line in the Pyrenees where even small glaciers may be found, is of an impassable sheer height, just notched enough at one point on the west to admit a painful scramble into the valley of the Gave d'Pau and on the east to admit one into the Valley of the Lys (into the basin of the Adour, that is) at one end, and into the basin of Garonne at the other.

A journey through Sobrarbe can be undertaken either from Sallent and Panticosa or from Gavarnie, and in either case your exploration of high Sobrarbe begins at the Hamlet of Bujaruelo, which the French call Boucharo.

How to reach Bujaruelo from Gavarnie I shall describe later : for the moment I propose a start on the Spanish side.

If you start from the Spanish side at Panticosa, a plain path takes you up the valley of the Caldares until you are right under the frontier ridge. There the path bifurcates ; you take the right hand branch along the chain of lakes that lies just under the wall of the main ridge, and you climb slowly up to the path at the head of it. The whole climb from Panticosa to this pass is 3040 feet, and it will take you from early morning until noon. Or, if you will start before a summer dawn, at any rate until the heat of the morning. For though it looks so short a distance on the map, and though there is no difficult passage, it is very hard going. The reason I mention this matter of hours is that when you have got down the other side into the valley of the Ara, you are still 8 miles by the mule

THE SUMMIT OF THE MARBORÉ

path from Bujaruelo, and though it is all down hill,
you will hardly do these 8 miles under two hours
and a half; however early you start, therefore,
the back of the day is likely to be broken by the
time you come to Bujaruelo. Once there a new
difficulty arises; for Bujaruelo is not a pleasant
place to sleep in. I have not myself slept there,
but the verdict is universal. Though you are
coming from a Spanish town the Customs may
bother you at this hamlet because they cannot
tell but that you have come over some one of the
high passages from France, such as the Pla Laube
up the valley. At any rate, unless you are going
to camp out you must push on to *Torla*, 5 miles
on down the valley, and you will pass through a
great gorge on your way. Now at Torla the
hospitality though large and vague, is good enough.

If, however, you are taking the Upper Sobrarbe
with the idea of camping, you must not go on
to Torla, but you must do as follows. Just at the
far end of the gorge of which I have spoken the
path crosses the river Ara by a bridge called the
Bridge of the Men of Navarre. There you will
see a path leaving yours to the left, and zigzaging
up the mountain side eastward. This is the one
you must take. It climbs 600 feet, gets you round
the cascade which here pours into the Ara from a
lateral valley, and finally puts you on to the level
floor of that lateral valley : it is called the valley
of Arazas. Here there is excellent camping ground
everywhere, and it will be high time to look for a
camp by the time you are well upon the floor of

that gorge; you may have to go up some little way to find wood, but much of this valley in its higher part is clothed with forests. The next day you must, as best you can, force your way to Bielsa, and unless the weather is fine you may very possibly have to sleep another night upon the mountain.

The trouble of this difficult bit is the great height of the lateral ridges. At the end of this fine valley of Arazas, which curves slowly up northward as you go, is the huge mass of the Mont Perdu, and you cannot get out of the valley without going over the shoulder of it. In order to do this proceed as follows and go along the stream until the path crosses over from the northern to the southern bank, at a place where the cliffs on either side come very close to the water. The path goes along under and partially upon the face of these cliffs in a perilous sort of way, until it comes to a lateral streamlet pouring right down the side of the terminal mountain. This lateral streamlet you must be sure to recognise, for upon your recognising it depends the success of your adventure; and you may know it thus: The place where your path strikes it, is exactly 1000 yards from the place where you crossed the main stream. When you come to this lateral streamlet you will see, or should see, a transverse path running very nearly due east and west; and up that in an eastward direction, immediately above you, a distance of 800 yards, upon the shoulder of the great mountain is the depression for which the path makes. It is called the *Col de Gaulis.*

For all of this by the way you will do well to
consult Schrader the whole time. What the going
is like on the further side of this col I cannot tell
for I have never come down it, but I know that
your way descends right by a very short and steep
gully in which a torrent makes straight for the
valley beneath, and I know that when you have
made that valley your troubles are over.

You fall through a descent of just under 2000
feet in a distance of less than a mile as the crow
flies. You must therefore be prepared for a very
steep bit of work. Once in the valley, however,
everything is straightforward. On reaching the main
stream of this new valley (which runs north and
south) you turn to the right, southward, and follow
its right bank between it and the cliff; you cross
a rivulet flowing from a deep lateral ravine about
a mile further on, and less than half a mile further
again see a new path leaving your path and going to
your left, crossing over the valley and its stream, and
making up a gulley which comes down facing you
from the opposing heights. Take this new path up
this gully (the path runs everywhere to the *south*
of the water), and you will find yourself after a
climb of somewhat over a 1000 feet on the Col
d'Escuain. Thence the way is perfectly clear,
running due south-west for 5 miles, just above the
edge of the cliffs of the gorge of Escuain, until you
reach the village of Escuain perched above that
ravine.

Whatever efforts you may have made and however early you may have started, you will hardly

have reached human beings again at this place until, as at Bujaruelo the day before, the back of the day is broken. Nevertheless, unless you are to camp out again upon the mountain, you must try and push on to Bielsa. It is more than 10 miles, however much you cut off the windings of the path, which takes you past the chapel of San Pablo, leaving the village of Rivella on the left up the mountain side, then across a steep cliff down to the profound gorge of the Cinca, from there an unmistakable road goes through Salinas de Sin and follows straight on up the valley to Bielsa just 4 miles further on.

If you can do that in one day you will have done well.

There is another and shorter crossing, which, though it is invariably used by the mountaineers I have not described because most people unused to the Pyrenees would shirk it. When you have come down from the Col de Gaulis into the valley below, if instead of going southward to the right you go northward to the left, crossing the stream, and climbing up on the further side of it, the path takes you at last to a very high col, called in Spanish the Col of Anisclo, but in French, the Col of Anicle. This col is not far short of 9000 feet high, and it is particularly painful to have to attempt it just after the difficult business of the Col de Gaulis. It means two ports within a few hours of each other, the second one 3000 feet above the valley, and what that is in the way of fatigue, a man must go through in order to know. Moreover, the descent on the far side from the Col of Anisclo is exceedingly steep.

However, if you do this short cut you have the advantage of finding yourself at once in the main valley of the Cinca and, when once you are on the banks of that river, you are not more than 8 miles or so from Bielsa by a good path leading all the way down the stream on the left bank. You save in this way quite 6 miles, and reduce your whole journey from the mouth of the valley of Arazas to Bielsa to a little less than 20 miles.

The distance you have to go before you come to human beings is much the same by either track. Escuain is just about as far from the Col de Gaulis, as is Las Cortez, the first hamlet in the Cinca valley. Again, by this shorter way you miss the gorge of the Escuain, but you see the huge cliffs of Pinède, which are perhaps the finest wall in the Pyrenees with their summits along the crest of 9000 feet, 5000 feet or more above the stream at their feet: it is the edge of this ridge of cliff which must be crossed at the Col of Anisclo. Either way therefore is as fine and either as deserted as the other. But the second much shorter and far more painful.

Before I leave this passage between the first and second of the Sobrarbe valleys—between the valley of Broto, that is (as they call the valley of the river Ara) and the valley of the Cinca—a few notes on the road should be added.

First, I have said that Torla, Bujaruelo (Boucharo) may be made from Gavarnie as well as from Panticosa. This is so ; and if you undertake the exploration of Sobrarbe from Gavarnie, it is a much easier business to get to Bujaruelo from the French hamlet, than it is to get to it from Panticosa.

The excellent road from Gavarnie to the top of the port is a very small matter, and from there down into Bujaruelo is an easy descent of three miles. If you start from Gavarnie, therefore, in the early morning, you can with an effort and in good weather go the whole length of the Val d'Arazas, over the Col de Gaulis, and the Col of Anisclo and sleep in Biesla that same night, or you can, taking it more easily, make a camp at the head of the Val d'Arazas, or you can break your journey in the valley between the two Cols of Gaulis and Anisclo, camping there for the night ; I am told the camping ground in this gorge is not very good, otherwise that would be the ideal place to break your journey.

You may next remark that in the lower part of the Val d'Arazas, right on the path, there is a good inn, which will save your camping out in the valley at all, if you are not so inclined ; but the inn is so far down the valley that it does not save you very much in the next day's walk. . Further, you should note that all this group of valleys, the Arazas, the Pinède (which is that through which the Cinca flows), the Velos, which is the stream at the foot of the Col de Gaulis, the Escuain, etc., are, unlike most others in the Pyrenees, true *ravines*. They correspond to what Western Americans mean when they use the Spanish word Cañons, that is *clefts*

sunk deep into the stuff of the world and bounded
by precipices upon either side. These not only
make the whole district a striking exception in the
Pyrenean range but also make the finding of and
keeping to a path necessary as it is throughout the
Pyrenees, more necessary here than anywhere else.
If, for instance, you lose the path at the head of the
Arazas, where it goes up the cliffs, you will never
make the Col de Gaulis though it is less than a mile
away, and if you miss the path up to the Col of
Anisclo you can never get down into the Pinède
at all.

It is worth remembering that from the foot of the
Col de Gaulis a path of sorts leads up the flank of
the mountain to the Spanish side of the Brèche de
Roland. I have never followed it, but I believe it
to be an easier approach than that over the glacier
upon the French side.

Once you are at Bielsa on the Cinca, you are in
the centre, and, as it were, in the geographical
capital of the high Sobrarbe, and it is your next
business to go on eastward into the last valley, that
of the Esera, the central town of which is Venasque.
Between the upper part of these two valleys and
right between these two towns lies the great mass of
the Posets, a huge mountain which lifts up in a
confused way like an Atlantic wave and is within a
very few feet of being the highest in the Pyrenees.
It is a mountain which though it is not remarkable
for precipices or for any striking sky line, should
by no means be crossed (though it can easily be
ascended), but must be turned.

The straight line from Bielsa to Venasque lies slightly south of east and is but 15 miles in length, but it runs right over the mass of the Posets and crosses that jumble of hills only a couple of miles south of the culminating peak. Venasque must therefore be reached by a divergence one way or the other, and one approaches it from Bielsa by going either to the north or to the south of the mountain group of the Posets. The northern way is a trifle shorter but much more difficult and much more lonely. On the other hand it takes one into the very heart of the highest Pyrenees, right under the least known and the most absolute part of the barrier which they make between France and Spain. I will therefore describe this northern way first, as I think most travellers who desire an acquaintance with the hills will take it.

From Bielsa a path going eastward crosses the Barrosa (at the confluence of which with the Cinca Bielsa is built) runs round the flank of the mountain and goes right up to the Col of the Cross " De La Cruz," 4000 feet above the town. You may know this pass, if you have a compass, by observing that it is due east of Bielsa. To be accurate, the dead line east and west from the top of the Col exactly strikes the northernmost houses of the town.

The eastern descent of the Col is quite easy and once down upon the banks of the Cinqueta, you see, half a mile to the north of you, the hospital or refuge of Gistain. From that point you follow up the valley north-eastward, on the right or northern bank of the stream under a steep hillside for a

FRONTIER RIDGE

COL DE Aigues Tortes

L DE SISTAIN 2524

1675

CAB. DE TURMO

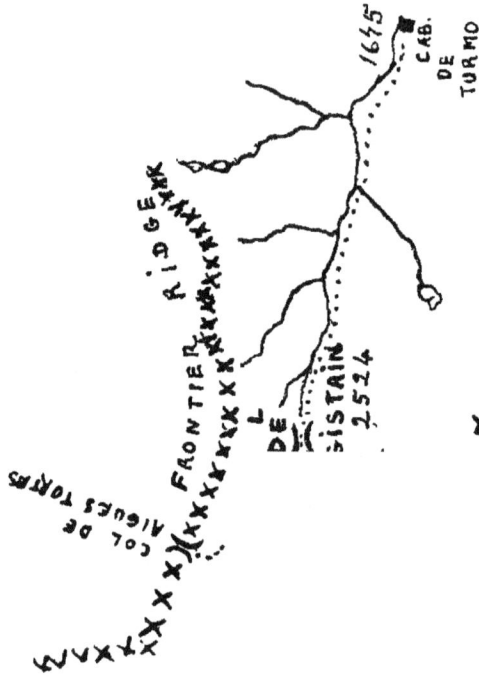

SUMM T OF POSETS 3367

Paths
Contours at 00 Metres

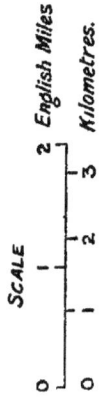

SCALE

0   1   2   English Miles

0   1   2   3   Kilometres.

FORK

N
W   E
S

Variation 14° W

COL DE LA CA 285

GIS

EL PL

couple of miles until you come to a fairly open place where the two upper forks of the Cinqueta meet. You cross the northern fork and go on eastward and northward up the eastern one, still keeping at the foot of the northern hillside.

What follows is not very easy to describe and should be carefully noted. What you have to pick out is a particular col on the opposite slope beyond the stream. This col is three miles or so from the fork, five from the Refuge, and is called "the Col de Gistain." As you go up this valley the opposing side is formed of the buttresses of the Posets. From that mountain four torrents descend to join the east fork of the Cinqueta, between the place where you crossed and the col you are seeking. The first torrent falls into the valley which you are climbing half a mile or so after you have crossed the north fork and begun the new valley; a second comes in about a thousand yards further on, a third about a mile further yet, and you may see each of them coming into the stream at your feet from down the opposing side, which consists, as I have said, in the buttresses of the Posets.

Another way of recognising these three torrents (and it is essential to recognise them) is to note that between the first and the second the slope is not violent, while between the second and the third it is a rocky ridge.

When you have seen the third come in, you must watch *exactly a mile further on* for the entry of the fourth. This fourth one is your mark by which to find the col. Just after passing in front of the mouth of this fourth torrent, your path, such as it is, will cross the Cinqueta, turn sharply eastward, and begin to climb up the right or northern bank of this fourth torrent.

The ascent is not steep, and in 1500 yards you are on the *Col de Gistain* between 8200 and 8300 feet above the sea, and almost exactly 3000 feet above the spot where you left the north fork of the Cinqueta to follow the eastern valley. Another way of making certain that you do not miss the all-important turning is to count the torrents coming in upon *your* side, the *north* side, of the valley; that is the torrents, each coming in from its own ravine, which your path crosses.

They also are three in number and fairly equidistant one from another, the first about a mile after you have crossed the north fork, the next a mile further on, and the next just under a mile beyond that. It is after you have crossed the third and have proceeded another 500 or 600 yards that your path to the Col de Gistain will go off opposite to the right, crossing the stream at your feet, and following the torrent that falls from that opposing side.

Yet another way of making sure is to watch (if the weather is fine) for the col itself, an unmistakable notch with a ridge of sharp rock just to the north of it and a less abrupt arète going south of it up to the summit of the Posets.

I have written at this length of the passage not only from the difficulty of discovering, but also from the danger that will attend any delay in finding it. If you go on past the turning where the path to the col goes off eastward you may get over the wrong port on to the French side, miles from anywhere, or you may take the rocks of the Anes Cruces and find yourself on a ridge beyond which there is no going down either way; while if you turn off too early you may climb right up on to the glacier of the Posets, and lose a day and be compelled to pass a night in that frost.

Once you have got to the top of the Col de Gistain, however, you are free. All the running water below you leads you down into the valley of Venasque; there is no steepness and no difficulty. The rudimentary path follows the stream, there is a little Cabane on the upper waters of it, soon the floor of the valley widens out a trifle, and four miles on, not quite 3000 feet below the pass, is another Cabane; that of the Turmo. The path from this point becomes more definite; it crosses the stream 2 miles down in order to avoid rocks upon the southern side, recrosses it again a mile later to negotiate a steep and narrow gorge, it comes over once again to the northern side by a bridge a few hundred yards further on, and almost immediately reaches the valley of the Esera at a point 9 miles or so from the summit of the pass. Here an ancient and remarkable bridge, the bridge of Cuberre, crosses the Esera, and enables you to gain the wide mule track to Venasque, which town lies rather more than 2 miles down the road.

It will be seen that the whole difficulty of this

passage lies in making certain of the Col de Gistain.

If I have exaggerated that difficulty I have fallen into an error on the right side, for to miss the col is to fail altogether and possibly to be in danger. If those who have approached the Col de Gistain from the east, or who have only seen the place in clear weather, imagine it to be discoverable under all circumstances, they are in error; indeed, if the weather is bad, it is just as well not to attempt the passage at all.

This northern way from Bielsa to Venasque is, as I have said, the most difficult. The southern way is as follows.

You go down the gorge to the Cinca by the road to Salinas de Sin, there the road branches, the main part goes on down the Cinca, the side road goes sharply off to the left up the first affluent of the Cinca, a lateral valley which points south-east, and is that of the Cinqueta. This road crosses the Cinca, follows the eastern or right bank of the lateral stream for some two-thirds of a mile, then crosses over and in about 3 miles from the crossing reaches the hamlet of Sarabillo. Thence it proceeds, still upon the same side of the stream and facing a considerable cliff upon the further bank, to the village of El Plan, which lies somewhat less than 5 miles up from Sarabillo, and is reached by crossing the stream again just before one comes to the village.

At El Plan one may repose. One will have walked by the mule paths more than 12 miles, and there is a long way before one.

The main path goes on to the next village, that of St Juan, and so up the Cinqueta to the hospital of Gistain where it joins the northern route we have just been tracing. The southern way, which I am now describing, is by a path leaving El Plan at the end of the village and going down to the river (which here runs through a broad valley floor), across the river by a bridge, and then up the torrent valley of the Sentina, a little south of east. The path runs on the right or northern bank of this torrent, and any path or tracks to be seen crossing the water are not to your purpose. Keep always to the same side of the stream until you come to the col, which is more than 4 but less than 5 miles from El Plan and is called the Col de Sahun. From this col the path continues a little less clearly marked, but quite easy, down the sharp valley on the further side to the village of Sahun which lies exactly due east of the col and just over 3 miles from it. The whole passage, therefore, from El Plan to Sahun, is a matter of not more than two hours, and from Sahun to Venasque there is an excellent mule road following up the open valley of the Esera; a distance of just 4 miles.

By this southern approach the whole distance is but a plain walk of under 20 miles with only one low and easy col to climb, but of course it tells you far less of what the Pyrenees can be than does the northern passage.

With the valley of the Esera and the town of Venasque you have come to the end of Sobrarbe, and of all that remote and ill-known district which is the most savage and the most alluring in these great hills. Indeed, you are no longer properly in the Sobrarbe, but rather in the subdivision of Ribagorza, which had a Count to itself in the Middle Ages, and was the march between Aragon and Catalonia. From Venasque you can get back again at your ease next day, by one of the best known mule tracks in the Pyrenees, to the French valleys and to wealth again at Luchon.

IV. The Tarbes Valleys and Luchon

THREE valleys, two profound, one shallow, depend upon and radiate from the town of Tarbes which stands in the plain below the mountains. Their rail system and their road system converge upon Tarbes, and it is from Tarbes that they should be explored.

The two long valleys are the valley of Lourdes, down which flows the Gave de Pau and the long valley of Arreau or Val d'Aure (it is the longest enclosed valley of the Pyrenees). The short valley is the valley of Bigorre, wherein the Adour arises.

For a man on foot these three valleys are of
interest chiefly in their highest portions alone.
The energy of French civilisation has penetrated
them everywhere with light railways and with roads,
and has united them all three by a great lateral road
running from Arreau to Luz over what used to be
the difficult and ill-known port of Tourmalet; while
it has thus done a great deal for those who only use
the road, it has hurt the district from the point
of view which I am taking in this division of my
book.

There is indeed one great hill which no development of roads can effect, and which is the chief
interest of all these three valleys for the man on foot.
It rises in the very centre of the district and is
called the Pic du Midi de Bigorre. This peak
stands thrust forward from the main range, a
matter of more than 10 miles from the watershed,
and isolated upon every side save where the isthmus
of the Tourmalet binds it to the general system not
much more than 2000 feet below its summit. But
the Pic du Midi de Bigorre, fine as it is, does not
afford so many opportunities to the man exploring
the Pyrenees on foot as do other peaks. It is a bare
mountain, all precipice upon the northern side, and
steep every way. There is no camping ground save
at the foot of it in the little wood above Abay.
Moreover, there is a road right up it, an observatory
upon the top, and arrangements for sleeping and for
eating and drinking as well. No other of the great
mountains of Europe have been put more thoroughly
in harness. The chief use of it (for the purposes

of this book) is that from its summit you will
get a better general view of the eastern Pyrenees
than from any other point reached with equal ease,
and that you can see in one view, as you look
southward, the Maladetta on your extreme left, the
Pic du 'Midi d'Ossau on your extreme right, each
about 30 to 40 miles away.   It is also a point from
which the sharp demarcation between the mountain
and the plain, which characterises the northern slope
of the Pyrenees, is very clear ; for this peak, jutting
out as it does from the mass of the hills, dominates
all the flat country beneath.

The roads of these three valleys are somewhat
overrun—even in their upper portions.   That from
the end of the light railway from Luz to Gavarnie,
is, in the summer, the only really spoilt piece of the
Pyrenees ; that from Arreau up to Vielle Aure in
the furthest valley is less frequented, but there is no
particular reason for stopping in it or for camping
in it, especially when one considers the waste spaces
on either side, where one may be wholly remote and
at peace.   There is, however, in one branch of this
valley, that is in the gulley which runs due south
from Trainzaygues, a good camping ground of
woods and stream.   A road runs up it to the refuge
of Riomajou at its summit, and from this two
difficult cols can be reached by two branch paths
which go over either shoulder of the Pic d'Ourdissettou, that on the right or west gets one down to
Real and Bielsa ; that on the left ultimately and
with some difficulty to Gistain and El Plan.   There
is also an entry from the main valley into the

Sobrarbe, going up the main valley through Aragnouet, and up the very steep pass called the Pass de Barroude ; one also comes out by this way on to Real and Bielsa, but it is by the other fork of the Spanish valley.

The pass called the Port de Bielsa proper marks what was once perhaps the main pass north and south over these hills. It leaves the valley at Leplan above Aragnouet and stands between the two passes just mentioned. These and all the difficult ports, springing from the three valleys of Tarbes and crossing the central part of the range, lead one into the Sobrarbe and the track described in the last division of this chapter.

The valley of Arreau has an eastern fork following the Louron at the head of which are further high passes, all in the neighbourhood of 8000 feet, which lead one into the Posets group and the eastern end of Sobrarbe. Of these the most interesting is the port of Aiguestoites, which is that upon which one comes by error if one misses the Col de Gistain on the northern way from Bielsa to Venasque.

The Cirques—the great semicircles of precipices —which have always been remarked as distinctive of the Pyrenees, are crowded in this region. The Cirque de Gavarnie is the most famous, and therefore, in our time at least, impossible for a man who really wants to wander. You cannot be alone there ; but the Cirque of Troumouse is not hackneyed and should be seen once at least. You may reach it by taking the road up from Luz to Gavarnie, and following it as far as Jedre. Here the Gave branches, you go up the zigzag of the road, past the church of Jedre, and take the path which leaves the highway to the left and follows up the eastern Gave, or Gave de Heas on its left bank. The path crosses that stream 2 miles further on and follows up the right bank to the little hamlet of Heas (which gives the torrent its name). It continues getting less distinct past the chapel of Heas; you turn a corner of a rock and find yourself in this huge, bare, deserted circle of precipices with the Pic de Gerbats at the left end of it, the Pic of Gabediou at the east end, and in the midst the highest point, the Pic d'Arrouye, which just misses 10,000 feet. The path continued will take you up past some Cabanes over the little glacier, and across that steep and very difficult ridge down into the Spanish valley of Pinede—which ends up, of course, in Bielsa.

But for these ramifications of their higher ravines, the three valleys of Tarbes are the least suitable for a man travelling on foot; of the three, however, the Val d'Aure will afford the most variety and the most isolation.

If, for any reason, one of these three valleys is chosen for a short holiday, Tarbes—where there is a good hotel, The Ambassadeurs—is the centre from which one should start and to which one should return; it faces right at the mountains, it is the most truly Pyrenean town of all the plain, and it is full of excellent entertainment. From Tarbes also start the three lines which take you up each valley, to Argelès, to Bagneres de Bigorre, and to Arreau.

Luchon

The valley of Luchon stands by itself as a separate division of the Pyrenees.   It has character altogether its own, formed both by political accidents, which separate it from its twin valley of the Upper Garonne —the Val d'Aran—and by its physical conformation which thrusts the level floor of it up further into the hills than any other of the Pyrenean gorges.   It is indeed made by nature to be one of the great international roads of Europe and to lead into Spain, for it resembles in many ways the trench running from Oloron southwards along which the main Roman road, and the main modern road, find their way into Aragon.   The valley of Luchon would undoubtedly have formed the platform for such a road had not two accidents interfered with that destiny : the first, the great height of the ridge at the end of this particular valley ; the second, the lack of open country to the south.

The Roman road from Oloron over the Somport finds a wide plain and an ancient city at Jaca, within a day's journey of the central summit.   But the valley of the Esera (which is the Spanish valley corresponding to that of Luchon) is a good three days travel in length before it gets one out of the hills, and the first town of the plains on the Spanish side (the modern symbol of whose importance is the presence of the railway) is Barbastro 60 miles in a straight line from the watershed, and not far short of 90 following the turns of the mule path and lower down the road which reaches it.

But for these accidents the way through Luchon would undoubtedly be the great avenue from Toulouse to Saragossa, and even as it is the pass over the ridge here (called the Port de Venasque) is the most trodden and the clearest of all the passes, other than those followed by direct highways.

The valley of Luchon is the very centre of the mountain system, for it lies just east of that division between the two halves of the mountains, the eastern and the western chains. It is a frontier also between two types of scenery and two kinds of travel. It is the last of the deep flat valleys running north and south, which are, so far eastward, the characteristic of the chain. Immediately beyond it, to the east, begins a combination of hills of which St Girons is the capital, and into which still further east penetrate the much larger valleys of the Ariège and of the Tet.

The Thermal Springs of Luchon, and a chance popularity which made it the wealthiest holiday place in all the mountains, have now fixed it as a sort of central spot which sums up all travel in the Pyrenees. For nearly a century it has had the character, which continually increases in it, of great luxury, and of a colony, as it were, of the main towns of Europe. But, for reasons which I mention when I come to speak of inns and hotels in these mountains, it is in some way saved from the odiousness which most cosmopolitan holiday places radiate around them like an evil smell. The influence of Paris is in some part responsible for better manners and greater dignity than such tourist places usually show.

The little town is very old; it is probably the site of the Baths which were mentioned as the most famous of the Pyrenean waters as early as the first century, and which certainly stood in this country of Comminges. For Luchon is the modern centre of the Comminges, and the Comminges is first historical district of the Pyrenees west of the old Roman Province.

For a man travelling on foot in the Pyrenees the chief value of Luchon lies in its being the only rail head which lies close against the highest peaks. Here one can have one's letters sent and one's luggage, and to this place one can always return from the wildest parts of the Sobrarbe, or of Catalonia, which lie on either hand just to the south-west and south-east. It is also the best place in the whole range in which to change English money.

The valley, though it has great historical interest (and everybody who has the leisure should see St Bertrand at the mouth of it), has, like those valleys to the west of it which have just been mentioned, little to arrest a man on foot, except in its last high reach. The ridge which runs north for 12 miles beyond Luchon and lies west of the railway, is high and densely wooded; but it is not good camping ground and it leads nowhere, while that to the east, less steep and not quite so densely wooded, has but one large field for camping, the forest of Marignac; and even in Marignac there is nothing but the wood to attract one. Once through the wood one is back again upon a high road and the valley of the Garonne.

THE RIDGE OF THE MALADETTA

Above Luchon, however, there spread out a number of valleys which are worthy of exploration in themselves, and one of which is the main way over into Spain. For this last we must continue the high road (which follows up the Pique, the river that waters all the Luchon district) until one comes, at the end of the causeway, to the hotel that was formerly a hospice, and is still called by that name. From this point a steep path takes one 3000 feet right up to the main ridge and to the little notch in the rock which is called the Port de Venasque. The path, though not so clear, is equally easy on the other side, bringing one down into the valley of the Esera and to the town of Venasque in the Sobrarbe. The whole way from Luchon to Venasque, counting this steep ridge, is one day's easy going. There is no way across the central range more simple or less difficult (though it is high), and it has very fine views ; as one crosses the summit one has right before one culminating peaks of the Pyrenees, the group of the Maldetta.

Just to the east of the Port de Venasque (which is about 8000 feet high—to be accurate, 7930)— is the Pic de Sauvegarde, a path which is almost a road leads up to it ; one pays a toll ; it is a sort of Piccadilly. The one purpose of the climb is to see from the summit a very good all-round view of the high peaks, which crowd round this turning point in the chain.

A less frequented valley, but one quite sufficiently frequented, is that of the Lys, which one turns into out of the main road by going off to the right, about

2½ miles after leaving Luchon, a carriage road, 4 miles in length, takes one up through the woods at Lys to an inn ; thence forward in the lovely valley and the half circle of peaks above, there is country wild enough for every one, but no good camping ground.

A further experiment for the man on foot, and one in which he will be more dependent upon himself and less in fear of invasion, is that of the Val Dastan, by which, and the high Port d'Oo, one can get down to Venasque. For this valley one goes up the new lateral road from Luchon as though one were going into the Val D'Aure and to Arreau. One may leave the road at any point after St Aventin to follow the stream below, but it is best to go on to a village called Gari, which is somewhat more than 5 miles from Luchon. At Gari is a road going south along a valley ; you follow that valley still going southward, till the road comes to an end in the neighbourhood of a wood which bars the upper end of the vale. A path, however, continues the line of the road, makes its way through the wood, and at the upper end of it you come out upon a fine lake. There is an inn to the south of this lake, and if you will go on a little north of the inn along the shores of the lake you will find very good camping ground. Indeed, it is wise to camp over night on this side of the range, for the climb up from Luchon is fatiguing, and the country of a sort inviting one to rest and look about one.

Rejoining the path it passes between two small lakes, just after leaving the wood, and climbs up the torrent past the little tarn called the Lac Glacé, immediately above which is the Port d'Oo. This port is a very high one, it falls little short of 9000 feet, and it is not more than a depression in the ridge around. On the further side a steep scramble marked by no path, gets one down into the valley beneath the Posets, and this valley is the same as that which I have described as lying to the east of the Col de Gistain and leading to the Bridge of Cuberre, and so to Venasque. It is a long and difficult way round to that town from Luchon by the Port d'Oo, but it is the wildest and therefore the best excursion one can make in the circuit of these hills.

I should mention before I leave this district that curious plain, Des Etangs "Of the Lakes," where is the Trou du Toro, a small circular pond.

The main source of the Garonne lies high up as befits the dignity of such a river in among the very noblest peaks of the Pyrenees; it springs from the eastern point of the Maladetta, flows down in a torrent to this plain "Of the Lakes," plunges into the little pond, and there wholly disappears! It reappears 2000 feet down at the Goueil de Jeou, on the northern side of the mountains, having burrowed right under the main range, and so runs down to Las Bordas. Sceptics to whom all in these bewitched mountains is abhorrent, from the realities of Lourdes to the legends of Charlemagne, annoyed by this miraculous action on the part of the Garonne, poured heavy dyes into the Trou du Toro, and then went and watched anxiously at Goueil de Jeou to see the coloured stream emerge ; but the Garonne was too dignified to oblige them, and the water came out limpid and pure ; as for the dye, it has stuck somewhere underground in the hills, and is colouring rocks that will never be seen until the consummation of all things at the end of the world.

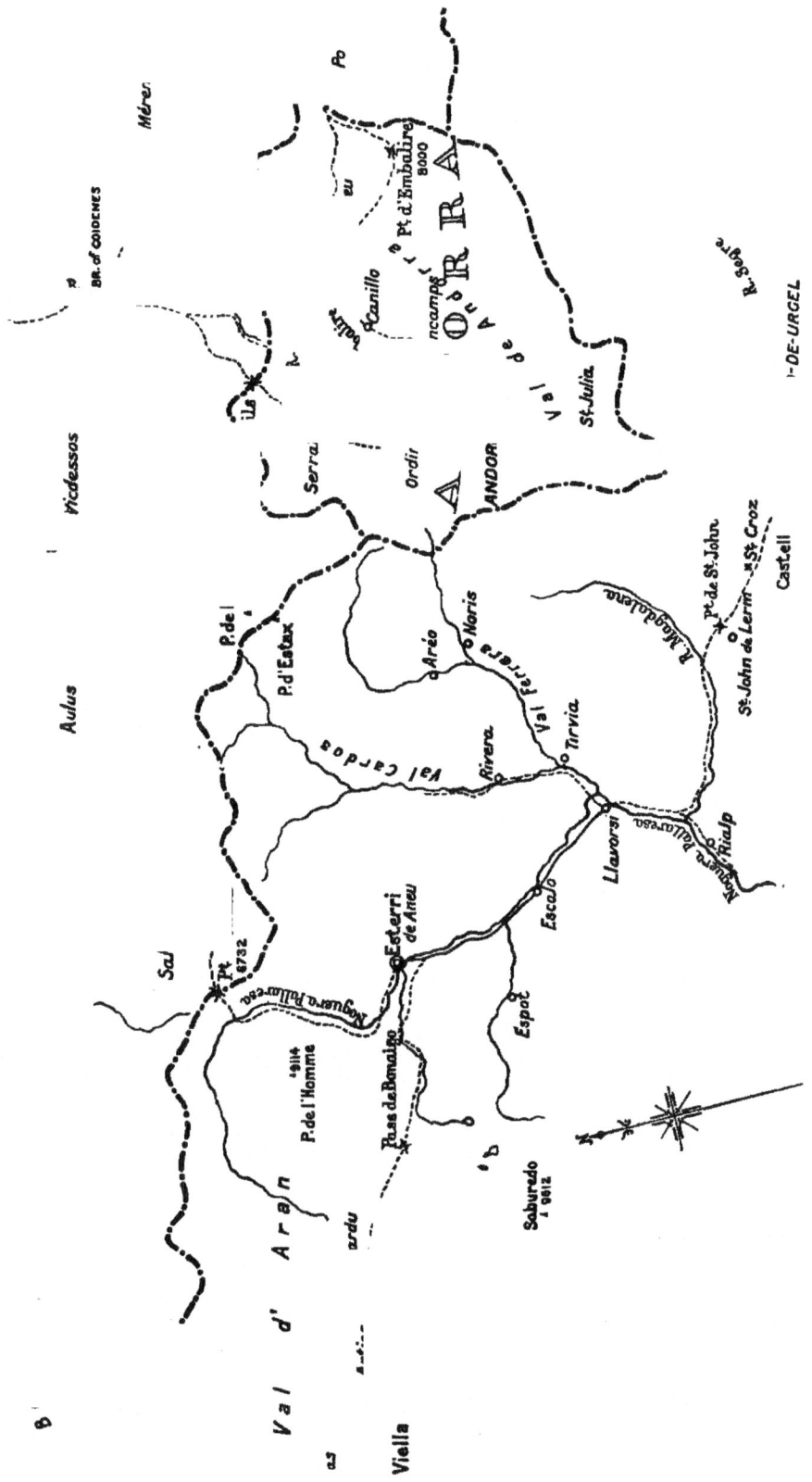

One may consider together Andorra in the Spanish valley of the Segre, the upper valley of the Noguera Pallaresa and Val d'Aran, for the journey through Andorra down to Seo, thence up out of the valley of the Segre into that of the Noguera, and so over to the Upper Garonne, makes one round, in which one covers one whole district of the Pyrenees, all Catalan.

There are two ways by which the curious country of Andorra can be reached from the north; both ultimately depend upon the valley of the Ariège.

The first shortest and most difficult way is by the vale of the Aston, a tributary of the Ariège which comes down a lateral valley and falls in near the railway station of Cabanes as the line from Foix to Ax; the second and easier way is by climbing to the sources of the Ariège itself, the main river, and over the Embalire.

As to the first—all the spreading rocky valleys which combine to feed the river Aston, form together a district of the very best for those who propose to explore but one corner of the Pyrenees during a short holiday. Even if such a traveller be unable or do not choose to force one of the entries into Andorra, he will have found on the Aston a country in which a man may camp and fish and climb anywhere, with a sense of liberty quite unknown in this kingdom. Here are half-a-dozen or more little lakes, deep forests, occasional Cabanes, good shelter, good bits of rock for such as like the risk, and outlines and distances of the most astonishing kind, and no land-lords. Of the many high valleys I have seen in the world, there is none less earthly than the last high reaches of the torrent which runs between the Pic de la Cabillere and the Pic de la Coumette, and which is the chief source of the Aston. The whole basin of this river includes six main streams, and, of course, many smaller torrents feeding these and the names of the peaks alone discover their desertion and the mixture of fear and attraction which they have had for the shepherds of these highland places. You may spend a week or a month or a whole summer in the neighbourhood and never come on this enchanted pocket which is bounded on the frontier by the high ridge running from the "silver fountain," the Fontargente, with its high peak and chain of lakes.

The Aston has at its sources, cutting them off from Spain, a ridge of 8000 to 9000 feet, it is a ridge the passes of which are but slight notches between the higher rocks.

The ways into Andorra across this ridge from the

Upper Aston are as numerous as these notches are, and nearly every notch can be climbed with knowledge and patience, but the only parts where something of a track exists are the Fontargente on the east, and the Peyregrils on the west. It is easy enough to fail at either, and there is therefore merit and sport enough in succeeding at either.

For the Peyregrils you must start from Cabanes and follow up the main stream of the Aston, by a clear path through the forest, taking with you the $\frac{1}{100,000}$ map as a guide. A little after a point where a bridge is thrown over the river, called the (Bridge of Coidenes) the two main streams of the Aston meet, one is seen flowing down from the south-east by the wooded gorge before one as one climbs, the other comes in cascades down a steep gully, pointing directly north and south. It is this gully which must be taken for the Peyregrils. One goes up over a steep rock still in the thick of the wood. On the far side of it one comes out into open grass country, and has one's first sight of the main range. The path comes down again to the stream, having turned the cascade, crosses the stream and flows along its right or eastern bank between the water and a range of cliffs which are those of the Pic du Col de Gas. About a mile from this crossing of the stream, as one goes on southward with a little west in one's direction, one comes to a side torrent falling in from the left; the path crosses this torrent, and still continues up the right bank of the main stream. It is a difficult point — for the path appears to bifurcate, and by taking the left hand branch, as I did four years ago, one may lose oneself in the empty valley under the Cabillere and be cut off for two days as I was, or for ever, as I was not. It is by making these easy mistakes that men do get cut off, and you may be certain that people who are found dead in the mountains under small precipices, are not, as the newspapers say, killed by some accident, but by exhaustion. They have wandered in a mist, or have been lost in some other fashion, until privation so weakens them that they no longer have a foothold; and in general, the great danger of mountains is not a danger of falling, but of getting cut off from men. Here, as in many other difficulties of this kind, your compass will save you; for if you find you are going more and more to the east, you are on the wrong path. The right one goes south by west along the left bank of the stream. There is a broad jasse or pasture which one traverses in all its length, one crosses another torrent coming in from a rocky gorge upon the left, the torrent and the path together turn more and more westward until one's general direction is due west, and at last one comes up against steep cliffs which are those of the Etang Blanc.

Thence, the way is plain, for the stream receives no further affluents and there is therefore no ambiguity of direction. The path follows the stream round a corner of rock whence one can see a tarn called the Etang de Soulauet, lying immediately under the water shed, and from that tarn the traveller goes straight up for 500 yards or so over the crest, straight down the steep further side, and

finds at the bottom of the valley the stream called
Rialb : such is the passage called the Peyregrils.

Once one is down on the banks of the Rialb, one
has but to follow the trail which runs along the bank
of that stream, cross it, reach the hamlet of Serrat,
and so follow the broadening water to the little town
of Ordino ; four miles beyond is Andorra the Old.
The whole distance from the pass to Andorra is
somewhat over 12 miles, counting all the windings
of the way. On this, as on so many crossings of
the Pyrenees, the difficulty is wholly on the French
side, once on the Spanish, the broader valleys lead
one without difficulty down one's way.

The other entry into Andorra from the valley of
the Aston, that by the Fontargente, is managed
thus :—

When the Aston divides just after the bridge,
one takes the south-eastern fork, one crosses the
bridge and finds a clear path going up the right
bank of ·the main stream of the Aston through a
wood. Four miles on this path brings one out of
the wood, and for another 4 miles it goes on still
following the same side of the stream in a direction
which is at first east of south, and at last curls
round due south. There is a bridge or two crossing to the other side, but one must not take them.
One must keep close to the eastern or right bank
of the Aston all the way until one comes to a place
difficult to recognise, and yet the recognition of
which is immediately essential to success. It is
a jasse rather narrow and small, lying between a
rocky ridge upon the left or east and a line of

cliffs upon the right or west. Here are a few Cabanes, and even if one has missed the place on first coming to it, it can be recognised from the fact, that, at the further end of this Jasse, the two sources of the Aston meet in almost one straight line, making with the main stream one has been following, a shape like the letter " T."

The path branches and takes either valley or arm of the " T "; it is that to the *left* or east down which one must turn—the one to the right or west leads nowhere but to the impassible cliffs and precipices of the Passade and the Cabillere. The eastern or right hand path then must be followed in a direction just south of east for exactly 1 mile, during all of which it keeps to the north of the stream At the end of that mile it crosses the stream, turns gradually round a high lump of rocky hill, going first south, then in a few yards south-west until it comes, at about a mile from the place where it crossed, upon the large tarn or small lake of Fontargente, " The Silver Water." The port lies in view just above the lake not 500 yards off. Once over it, it is the same story as the Peyregrils, a trail following running water which leads one through the upper villages to Canillo, the first town, to Encamps, the second one, and so down to Andorra the Old. The distance from the main range to Andorra by this trail is 2 or 3 miles greater than by the Peyregrils.

These are the two difficult and mountain ways of making Andorra from the north.

The easier and much the commoner way is

to approach it from the upper waters of the Ariège.

One takes the main road from Ax to Hospitalet up which there is a public carriage or "diligence"; it is as well to go on foot, for one will get to Hospitalet before the diligence if one starts at the dawn of a summer's day, and it is important to get there early as there is no good sleeping place between the French side and the town of Andorra itself. At Hospitalet the main track for Andorra runs down in a few feet to the torrent of the Ariège, crosses it, and follows its left bank. It goes over the frontier which is here an artificial line, and though you are still on the French side of the range, you are politically in Andorra, upon this deserted grassy slope which forms the left bank of the Ariège.

At the second torrent which comes down this slope into the river—or rather the second stream, for they are quite small—the telegraph wire, which has hitherto followed the path, will be seen going over to the right, up a somewhat steep side valley. This is at a point about 4 miles from Hospitalet. You have but to follow that line if it is fine weather, and you will come right over the ridge and down on to the Spanish side of the Andorran Hamlet, Saldeu. If it is misty on the heights you will almost certainly lose the line, and possibly your life as well. Nevertheless the crossing can be made even in bad weather by going somewhat further south to the point called the Port d'Embalire. To find this needs a certain care.

Note with your compass the trend of the Ariège; it curves round more and more as you follow it, and when it begins to point *due south* (which it does after a perceptible bend) you may note a fairly plain track coming down from the opposite side of the valley: it comes down and strikes the Ariège at a spot almost exactly 2 miles from the place where the line of the telegraph left the stream. Here opposite the road turn sharp up away from the Ariège (which is now but a tiny brook) and go *due west* by your compass right up the mountain, which is here nothing but a steep grassy slope, and you will strike the Embalire.

It is one of the few crossings which can be made in any weather, because you will find upon that slope, a little way up, the beginnings of a made road; that road was never completed. It has never been metalled, but it is culverted and graded, and is as good a guide as the best highway in the Pyrenees could be. Probably it never will be finished, for the Andorrans are opposed to an easy entry into their country; but so long as its platform remains, one can never lose one's way upon the Port d'Embalire. The further side is a steep and easy descent over a sort of down, and one finds Saldeu by this longer route about 4 miles from the summit. Whether one has followed the telegraph line or come over by the Embalire, the two tracks join at Saldeu, and the rest of the way is identical with that which you will come to by Fontargente, that is, through Canillo and Encamps to Andorra the Old.

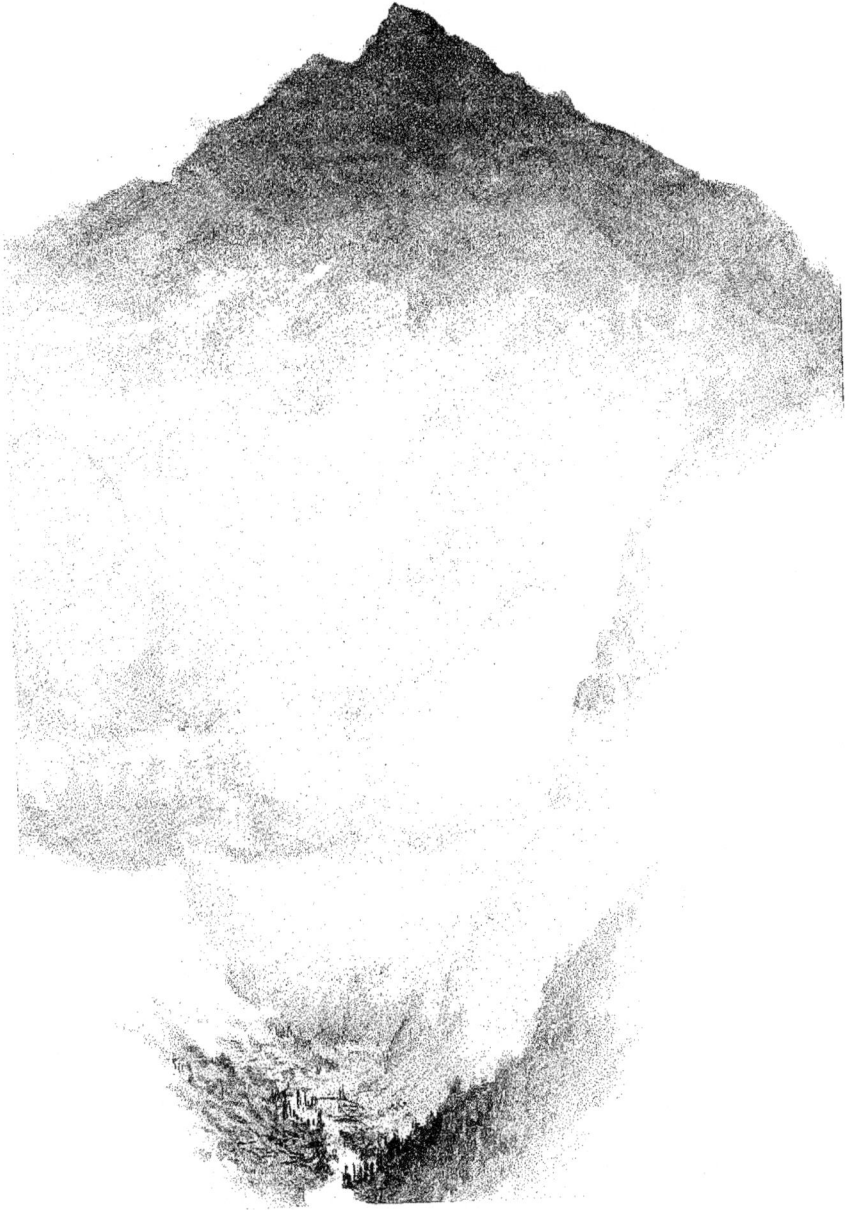

THE EMBALIRE FROM THE SPANISH SIDE

Easy as the way is, however, it should be remembered that it is a long day from Ax, for counting every turning, it is not far short of 30 miles, and
more than half of that is up hill. Ax stands at
about 2000 to 2400 feet (according to the part of
the steep town one measures from) and the summit
of the Embalire is almost exactly 8000 feet.
There is no break in the rise from one to the other.

The interest of Andorra lies in its survival, and
the recognition it receives of being an Independent
European State. All these enclosed valleys of the
Pyrenees led a more or less independent life for
centuries; from a decline of the Roman Power
until the union of Aragon and Castille on the
Spanish side, and on the French side in some
places, up to the Revolution itself, they boasted
their own customs and could plead their own law

The violent quarrel between Madrid and Aragon,
in which the independence of Aragon was fiercely
destroyed, affected the greater part of the Spanish
Valleys, and killed their independence; but it
did not attack the Catalan valleys — of which
Andorra was the most secluded and remote, and
therefore Andorra survives.

One may study in Andorra what all these valleys
were in the long period of local and natural growths
between the very slow death of the Roman Bureaucracy, and the rapid rise of the modern. The
French, through the Prefect of the Ariège (as
representing the Crown of France, which in its
turn inherited from the county of Foix) claim a
partial control over the Andorrans who pay to the

Government in Paris £40 a year in fealty. The Spaniards have a hold on it through the Bishop of Urgel, who is not only their Ordinary but also their Civil Suzerain : he gets only £18 a year from the embattled farmers.

The Andorrans have all the vices and virtues of democracy clearly apparent. They are very well to do, a little hard, avaricious, courteous, fond of smuggling, and jealous of interference. Also in Andorra itself one great shop supplies their external needs, and conducts all their international exchanges. Catalan, a provincial dialect in Spain, is here the national language. They are divided, as are all Catholics, into Clericals and Anti-Clericals, the Clericals making, I believe, a working majority, and there is not among them, so far as one can see, a poor man or an oppressed one.

From Andorra the Old, a good open path leads through the narrow gates of the country, down on to the valley of the Segre, and so to Seo de Urgel.

Though it is but a few hours' walk from Andorra to Urgel, it is as well to pass the remainder of the day and the night at Urgel, especially if it is the first Spanish town you have seen, as it is the first for many people who cross the mountains at this place. You will certainly find nothing more Spanish along the whole range. This lump of a town with its narrow oriental streets was the pivot of the Christian advance into Catalonia. The Carolingian armies came pouring through that easiest of the passes, the Cerdagne, enfranchised Urgel, first of all the Mozarabic Bishoprics, and may be said to have refounded its Christian existence. For some reason difficult to discover Urgel fossilised quite early in the Middle Ages. No line of travel, no road linked up the long valley of the Segre, the armies and the embassies of the French knew nothing of Lerida, and it is characteristic of Urgel to-day that even to-day there should be no great road beyond it up the valley.

From Urgel your road back into France through the upper valley of the Noguera Pallaresa, and the Val d'Aran is difficult to discover in its earlier part, unmistakable in the high mountains; which is the reverse of the rule usual in other crossings of the hills.

You must go down the high road which runs south of Urgel until you come, in something over a mile, to Ciudad, which is that hill-pile of white houses, once fortified, which rises over against the Cathedral city.

There you must ask the way to Castellbo, which is two or three hours away up a torrent bed, and you must go up this torrent bed by way of a road.

If you start early from Urgel you will be at Castellbo well before noon, and the hospitality of the place is so great that you will wish to stay there. There is only one drawback to eating at Castellbo which is that you have after it to make a passage of the mountains which, though here not very high, well wooded and fairly inhabited, do not bring you to proper food and shelter until you have gone close on 20 miles and have reached Llavorsi in the further valley of the Noguera; and so, if you stop to eat your mid-day meal at Castellbo, it is quite on the cards that you will have to camp out in the hills and that you will not make Llavorsi until noon of the following day; for the col in between, though it is very easy, is higher above the sea than the Somport.

From Castellbo you have but to ask for the village of St Croz, which is perched upon a height just up the same valley, but from there to the port the way is difficult to find for the very reason that there are no *physical* difficulties. It is all one long ridge of wooded grass like a down, with rather higher peaks to the right and to the left and with more than one indication of a path several directions. A good rule, however, for finding the exact place where you should cross, is to make for a spot due north-west from the village of St Croz, and this spot is further distinguished by the fact that it is on the whole lowest upon the whole saddle. It is a mile and a quarter or a mile and a half from the village, and as you go to it over the easy grass you get a superb vision of the Sierra del Cadi barring your view of Catalonia and standing up against you much higher than ever it seemed from the floor of the Cerdagne. No hills in Europe look so marvellously high.

As the saddle of this port, which is called the port of St John, is so long and easy it might seem indifferent at what point one crossed it; it is on the contrary very important to get the *exact* place and for this reason, that on the further or north-western side of it there is a profound ravine densely wooded, if one does not make the *exact* spot one has no path through this wood. That means hours of delay and one may very well come out upon the right instead of the left bank of the ravine; in which case in order to find the trail for Llavorsi at the bottom of the valley one may have a precipitous descent into the ravine and a bad climb out of it on the other side. Look, therefore, carefully for the path which begins to be clearly marked the moment the saddle is crossed, and follow down it until you come to a steep rock which overhangs the main stream at the bottom of the valley. This main stream is the Magdalena and runs not quite 2000 feet below the summit of the port. The trail is very distinct when once one has reached the valley; small villages are passed; it climbs up on the left bank to avoid a precipitous place and comes down to the water again at a place where the Magdalena falls into the main stream of the Noguera.

Here you must descend to the floor of the valley and take the road which is being made and which will in a few years form another great international highway up the valley of the Noguera. The road runs all the way on the left or eastern bank of the stream, which is broad and rapid and confined by very high steep hills upon either side. Three miles from the place where the path descended to the junction of the Magdalena and the Noguera, you will find another large river coming in. The road crosses by a wooden cantilever bridge where one pays a toll (I think of ½d.), and once across one is in the unpleasing village of Llavorsi.

The valley opens somewhat and is called Aneu, having on the left the exceedingly rugged and tangled chain of the Encantados, a wilderness of rocky peaks and lakes—and on the right a clear ridge which cuts off this countryside from the Val Cardos and the Val Farreira, both wild districts at whose summits is a bit of country as lonely as the Upper Aston.

All the way from Llavorsi up this Aneu valley the new road runs. I have not visited it for four years, and by this time it must be nearly finished, at any rate it is perfectly straight going and in all between 10 and 12 miles, with the exceedingly filthy village of Escaló about half-way.

It is not easy to give advice about sleeping in this walk from Urgel to Esterri. The distance between the two towns in a straight line is less than thirty miles, but the perpetual turning of the path makes it quite forty by the time one has reached Esterri, and what with the casting about for the right crossing on the port and the height of that crossing, it is too much for anyone to try and do in one day. Even if one were to sleep at Castellbo it would not mend matters much, for Castellbo is but a sixth of the distance, if that, and I would not recommend sleeping at Llavorsi. I have said that if one ate at Castellbo in the morning, it would mean camping out in the woods below the port of St John and this is perhaps the best plan after all: to leave Urgel on the morning of one day, to camp in the deep woods above the Magdalena and to sleep at Esterri, on the night of the second day. There is a good inn at Esterri, where everything is comfortable and clean, and the whole place is more civilised than any other town or village in the Pallars.

The next day you will go over the Pass of Bonaigo into the Val d'Aran, unless you prefer the much less amusing walk by the new road up over the Port de Salau to St Girons. It is less amusing because it gets you into France almost at once, whereas the walk into the Val d'Aran keeps you in Spain and shows you a very interesting geographical and political accident of the Pyrenees.

The town of the Val d'Aran is called Viella, and it lies 20 miles west by north of Esterri, between the two there is no obstacle but a high grassy saddle called the Port of Bonaigo the summit of which is exactly 3283 feet above the floor of the Noguera at Esterri, and the interest of which lies in this, that it stands right upon the junction of that "fault" which was mentioned in the first division of this book.

The Bonaigo is the exact centre of the Pyrenean system   On your left as you cross it, to the south that is, is the Saburedo, which is the last peak of the western branch. To your right upon the north the hills lift up to the Pic de l'Homme, which is the terminal peak of the eastern branch; and the ridge uniting these two branches runs in a serpentine fashion north and south with the saddle of the Bonaigo for its lowest point.

18

You will reach the summit, going easy from Esterri, in about three hours, and thence you will see, if the weather is clear, the distant snow of the Maladetta to the west, and in the vale at your feet, the first trickling of the Garonne. For by the twist the watershed here takes, you are crossing geographically from Spain into France, though the valley of the Garonne before you is still politically Spanish. The descent upon the Val d'Aran is somewhat steeper than the ascent from the Noguera, a path of sorts begins at the foot of it, and runs down the Garonne to the first hamlet, the name of which is Salardú. At Arties, a road begins, and 5 miles further on you come to Viella and to rest.

In Viella there is nothing but oddity to note : the oddity of a French valley governed by Spain. You are quite cut off, you will hear no news, and the only sign that you are on the north of the mountains will be the great and excellently engineered road leading down the Garonne from gorge to gorge, and reaching at last the French frontier at a narrow gate where is the " King's Bridge." Some miles further on is the French railway-head at Marignac. An omnibus starts in the early morning from Viella at whatever hour it pleases and gets down to the French railway in time for the mid-day train, but whether you take it or walk down on foot, you had better stop at Bosost, not half-way down, and there take the whittle woodland road eastward over the frontier by a very low gap called the Portillon and so saunter into Bagnères de Luchon, the noisy and wealthy capital of luxury. To come into Luchon suddenly after such a journey is as sharp a change as you can experience perhaps in all Europe. Do not forget before you reach Bosost to look up the gully which comes in from the left at a place called Las Bordas, some six or seven miles from Viella. This gully is that of the true Garonne, the fork of the river which we saw having such strange adventures rising on the wrong side of the main watershed of the mountains, burrowing right through them in a tunnel and coming out upon the northern side ; surely the only river in the world which behaves in such a fashion.

The walk which I have just described will have shown you most thoroughly all the wild north-western corner of Catalonia, and have taught you Andorra as well. Whether you take Cabanes for your starting place, entering Andorra by the difficult passes of the Aston, or whether you take Ax for your starting place and enter by the easy pass of Embalire, you will not make the whole round to Luchon in the best of weather under six days, and indeed a man who has but a week in which to begin to learn the Pyrenees, might very well choose this little square of them for his first introduction.

VI. CERDAGNE

THE Cerdagne forms a district quite separate from the rest of the Pyrenees. Its scenery differs from that of the rest of the range, its facilities for travel, its politics, everything in the place is different ; and though both valleys are Catalan, it is well not to include in the same summary a description of the Cerdagne and a description of the Rousillon.

The Cerdagne is the only broad valley in the Pyrenees, and it is a broad valley held in by walls of high mountains. All the other trenches which nature has cut into the range, are, without exception, profound and narrow. They expand occasionally into enclosed circles of flat land, the floors of ancient lakes, with a circle of steep banks all around, first wooded, then rocky, and reaching almost to Heaven. But these solemn circuses of secluded land, held in by narrow gates at either end, and small compared with the rocks around them, have a totally different effect upon the mind from those produced by such a landscape as the Cerdagne. You here have a whole country side as broad as a small English county might be, full of fields, and large enough to take

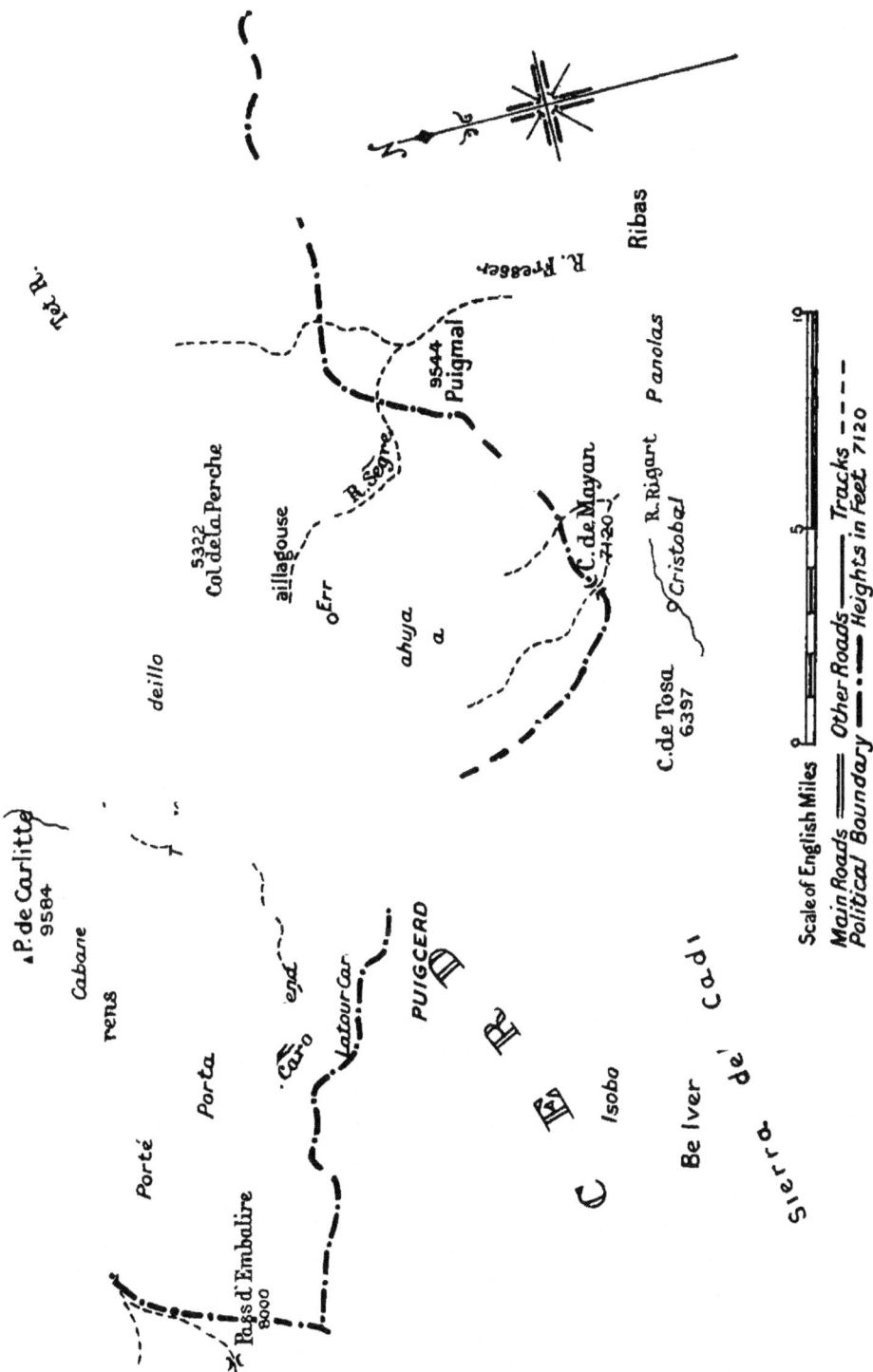

abreast a whole series of market towns. This is the sort of plain, which, were it bounded by hills, rather low like our English downs, would seem a little country by itself: a place large enough to make up one of our European divisions, like the countries of England, or the minor provinces of France. A broad river valley, such as decides a score of places scattered over Western Europe, here binds many households all united historically and defines a corporate condition for a fixed community of men.

This picture is framed in two great lines of hills roughly parallel to each other, and the effect when one comes upon it out of the last of the narrow valleys, may be compared to the effect upon a child's mind when he first sees the sea.

In order to perceive the full contrast of this exception in the Pyrenean group, it is best to approach it from the west ; whether you are coming on foot over the foot hills of the Carlitte groups down on to Mont Louis or Targasonne, or whether you are coming by the high road over the pass of Porté, there comes a point in your journey where, after so many gorges and narrow cliffs, the hills here suddenly cease at your feet and you see the whole sweep of the Cerdagne as broad as a field of corn ; you will have seen nothing like it all your way from the first foot hills of the Basque and the shores of the Atlantic.

On the eastern side, beyond the plain, you see the long ridge which is among the highest of the Pyrenees, and which stands steeply out of the flat. It stretches, as it were, indefinitely away into Spain and was called for centuries by the Mohammedans, and still is, the Sierra del Cadi. At its feet are a group of villages and towns, Saillagouse, Odeillo, Bourg Madame, Puigcerda (with its curious little isolated hill), Angoustrine, Palau, Osseja, Nahija, Err, and Caldegas, and that fascinating territory Llivia, which stands enclosed, making a little island of Spanish territory in the midst of French.

The structure of the Cerdagne explains its history. It is a slightly sloping shelf upon the Spanish side of the watershed, but the watershed here is not as it is everywhere else a steep ridge with rocks, it is a large imperceptible flat which, for the first few miles upon the northern side, slopes quite gently down towards the valley of the Tet, and on the south side slopes still more gently and easily away towards Spain. The Segre, the last and largest tributary of the Ebro, rises in this gentle plain in innumerable rivulets, which joins innumerable other rivulets at Llivia, and then receives the river of Val Carol, the river of Angoustrine, and the little river of Flavanara below Puigcerdá. There is in the whole extent of this plain no natural feature to form a frontier, and (as its upper waters form the only approach to the province of Rousillon) Mazarin, when the treaty of the Pyrenees submitted the Rousillon to the French Crown, claimed as a sort of right of way, the upper stretch of this wide plain.

The negotiations were not difficult, the frontier was drawn just so as to give the French Government everywhere the road down the Val Carol and up by

THE WALL OF THE CERDAGNE

Mont Louis to Perpignan. It was not the frontier between two civilisations or languages, the few square miles of the French Cerdagne, which is geographically Spanish, are Spanish also, Catalan Spanish, in customs, hours, architecture, and even cooking. It is Spanish in everything save the functions of government; and here you see just what differences government can and cannot make in a countryside. Government, where it exists against the will of the governed, effects nothing; but here there is no such friction, and you may compare the contented Cerdagne, which takes its orders from Paris, with the contented Cerdagne that takes them from Barcelona and Madrid. The subtle effect of the contrast is sufficiently striking; it is seen in the type of roadway, the paving of courtyards, in clocks that keep time upon one side and not upon the other, and in a certain hardness, which French assurance breeds, and which the Spanish ease avoids. It is a good plan as one enters the Cerdagne to take the by-road which leads straight across the plain from Urgel to Saillagouse. This by-road, when you have pursued it for about a mile, enters the isolated Spanish district of Llivia, and when you reach that town you find yourself in Spain, although all the villages round you in a circle are French villages. You have the Spanish delay, the Spanish tenacity, and the Spanish disorder. On coming out of it again, and immediately over the stream on the first village, the influence of the distant prefecture and of a strong hand upon the local community is apparent.

The Cerdagne has one bad drawback that, for all
its beauty and wealth, its entertainment is bad.
There is not, I think, one good inn in the whole of
it, and at Saillagouse, where the exterior looks
most promising, the people are so hard-hearted that
there is no comfort to be found under their roofs.
If you are thinking of food, the best place perhaps
for your headquarters is the little village of La
Tour Carol.   But if you are thinking of sights, your
best headquarters is the town of Puigcerdá, just
beyond the Spanish frontier, 3 miles or so from
Latour.

Puigcerda is the capital of the Cerdagne, and
there the people gather as to a fair.   It was the
capital of the Cerdagne long before the people knew
or cared whether they were governed from the north
or from the south.   One and a half miles away,
over the river in French territory, the tiny hamlet
of Hix marks the place where the old capital was
before Puygcerdá was founded and ousted it in the
early Middle Ages.   From many points in Puigcerda,
from the terrace in front of the Town Hall, from the
northern end of one of its streets, but especially
from its church tower, you take in one view the
whole of the Cerdagne.   As one gazes upon that
view, one should remember that this was the
principal highway of organised Christendom against
the Mohammedan, and through this went Charlemagne and his son.

The Carolingian tradition is nowhere stronger,
strong as it is throughout the Pyrenees, than in this
fruitful plain.   The very mountains perpetuate it

with the name Carlitte, and the valley of Carol and
the popular songs perpetuate it also.   It was this
broad floor, full of provisions and free from am-
buscade that allowed Christendom to dominate
Catalonia, and render free the country of Barcelona,
first of all Spanish territory, from the weight of
unchristian government.   It is the Cerdagne, therefore, to which we owe the later segregation of
the Catalonians from the rest of Spain, their forget-
fulness of warfare, their active commercial unrest,
their modern submission to Jews, their great wealth.
The Cerdagne should possess a great road throughout, for it is all of one type and all one valley.
By some historical accident it is not yet (I believe)
so served throughout.   After Puigcerda there is a
passage for wheeled vehicles for one-third, but not
for one half of the way to Urgel.   Belver, when I
last asked, was the head of the road, thenceforward
one must trust to a mule track.   I am speaking of
a part of the valley which I have not myself
traversed, for I only know Urgel from the other
side, and roads are being built so rapidly in the
Pyrenees that one cannot be certain of one's
accuracy in such a matter, but I believe there is not
yet any straight way to Urgel from the French
Cerdagne.   The only highroad from Puigcerda
turns out of the valley of the Segre and runs off
south and east to Barcelona.   Yet certainly Urgel—
that town we spoke of in connection with Andorra—
everyone travelling in this part should see: Seo,
the " Bishopric," the " See "; a sort of Bastion first
thrown out against the Mohammedans by Charle-

magne. It is more intensely Spanish perhaps than any other large town in these hills, and that because it is so thoroughly cut off from communication with the north. Here also you can find good hospitality. The people are kind, and local travellers are common. Urgel is, however, more easily approached from Andorra than from Puigcerda. And upon that account I dealt with it in connection with the little republic.

THE ARIEGE & TET VALLEYS

Scale of English Miles

Heights in Feet .... 9199.
Railways ....
Main Roads ....
Other do. ....
Tracks ....

PRADES

Fuilla

Ville franche

Oiette

Railleu

Tet R.

Sauto

Mont Louis

Fourmiguères

C. de Creu

Matemale

Espousouille

Angles

Riolb

R. Aude

Tet R.

Querigut

Rouze

Porteille
▲8243

Tet R.

P. Prigue
9199

P. de Tarbesou
▲8762

The Forges

L. Nougille

L. Roa & L. Torte

L. Lanoux

P. de Carlitte
▲9584

Orgeix Orleu

Ortege R.

Gnoles

Cabane

Merens

P. d'Auriel
8835

Beraque

Pass de
Puymorens

Porte

Hospitalet

Ariege R.

VII. The Tet and Ariège

THE valley of the Ariège is a basis for going either southward into Andorra by the tributary valley of the Aston or westward into Rousillon around the flanks of the Carlitte. Of the former journey I have spoken in connection with Catalonia The latter takes one into the valley of the Tet, and so to the Canigou which is the principal mountain of that valley. The highroad up the Ariège and over the Puymorens Pass into the Cerdagne and so into the Rousillon does not concern us here. It is designed for travel upon wheels. For going on foot the district is concerned with the Carlitte and the Canigou.

If one means to spend sometime in the big group of the Carlitte, one's headquarters must be Porté, the little village just over the Puymorens Pass. It is from here that the ascent of the highest peak is made and from here the fishermen start for the lakes that surround that Peak. If, then, one proposes to spend some days camping in the mountain and going nowhere in particular, it is from Porté that one must start, as the nearest point to the summits. On the other hand, nothing

can be bought at Porté nor for miles around, and
if one ascends the mountain from Ax, though the
distance is greater, one is more in touch with
provisions.

The Carlitte Group is remarkable for the number
of lakes, some quite large, which are to be found in
the hollows just under its highest ridges. On the
north is the large Lake of Noguille with the two
little tarns of Rou and Torte just above it on one
side; on the other, two little tarns lie under the
Pic d'Ariel. The main Lake is 6000 feet above
the sea, not far short of a mile long, 500 or
600 yards across, and very little visited. On the
south of the highest ridge and to the east of the
summit of the Carlitte, just above Porté, lies the
still larger lake of Lamoux. A good mile and a
half in length, but narrower than its twin upon the
north. Besides these two is the little group of
lakes at the source of the Tet, another group at
the sources of the Ariège, and another of half a
dozen and more just under the eastern cliffs of the
Carlitte which feed the big marsh of the Puillouse.

Unfortunately all this district, which is so wild
and open for travel, and so full of good fishing, has
but few camping grounds. The forest on the east
of the Carlitte is one of the largest in the Pyrenees,
and one may camp anywhere within it; but for a
lake as well as wood one can find but four spots:
one, the Camporeils; the other, the little pond just
above Langles; the third, a whole group of lakes
a mile south and a little west of the marsh of
Puillouse. It is by these last that one will do well

to camp if one is making one's way over the mountain eastward to Mont Louis, for they are within 5 miles of that town, and just beyond it is the valley of the Tet. The best camping ground in the neighbourhood of Ax is the fourth spot, at the Northern end of the lake of Noguille. Here the lake, the stream flowing from it, and the wood are all close together and as good a camping ground as any in these mountains can be chosen. The way to reach this is to leave Ax by the western road which branches off from the great national road and runs up the valley of the Oriège to Orgeix. Beyond this little village of Orgeix is another little village, Orleu, and beyond that again at the head of the highroad and not quite 5 miles from Ax is the point where you must turn off for the lake. It is not easy to find because the whole distance is very similar for miles. I will describe the way as best I can.

After the road leaves Orleu you have upon the left very precipitous steeps, rising to a height of some 6000 feet (or more than 3000 above the dale) covered with a forest which comes down very nearly to the road. On the right is a stream, and beyond it another belt of wood, less steep, with bare and high rocks above Somewhat over an English mile, from the Church of Orleu, a path leaves the road to the right and crosses the stream, taking its way upwards through the opposing wood; this path will lead you to the lake, but it is not the best way. The best way is to go on further, somewhat over half a mile to a group of huts called " The

Forges." Here you will see on the other side of the stream a valley running towards you from the mountain and coming from due south as you look up it. This valley, or rather ravine, is that of the torrent called Gnoles, and this is the gully you must follow. It falls into the Oriège just by the forges. You must go some yards beyond this junction of the streams and a path will be seen going right off at a right angle to the road and making for the gulley opposite. It crosses the Oriège at once, crosses the torrent almost immediately after, climbs up the steep on its left bank, crosses again on its right bank, and thence keeps on due south between the rocks and the stream, through the wood, until, at a point the height of which I cannot discover but well over 2000 feet above the road, it comes out suddenly upon the lake.

Here is the best camping ground within a reasonable distance of provisions and succour, and yet quite remote enough for a hermit. Here with the aid of the $\frac{1}{100,000}$ map, one may wander and take one's luck in the whole of this district of high peaks, rocks, and tarns, which stretch every way for 8 or 10 miles around.

If one's object is to make one's way into the valley of the Tet, instead of spending one's time in the mountains, the direction is straight and the way apparently easy, but it contains one difficult passage.

Your business is to make from Ax to the village of Formiguères, which is politically in the Roussillon, and lies south-east by a trifle east from Ax, and,

as the crow flies, barely more than 15 miles
away. You will, however, hardly get there under
20 miles of going, and it is unlikely that you
will do it in one day.

The first part of the road is plain enough. You
follow up the valley of the Oriège, as though you
were going to the lake of which I have spoken, but
instead of crossing over at the forges and going
south towards the lake, you go straight on up the
valley. Your path is not always distinct, but your
main direction is to stick to the Oriège as it gets
smaller and smaller in the high valley, and to look
out for a path which runs along that stream on its
left or southern bank.

For about 4 miles from the Forges you continue
climbing up the high valley of the Oriège, which
is wooded upon either slope, until you come to a
place where the wood recedes upon either side
(though there is wood in front of you), and the path
crosses the torrent to the opposite or right bank.
It is here that the difficulty of the way begins.

The path, you will notice by your compass, is at
this point going due south, for the Oriège has curled
round in that direction. Five hundred yards in front
of you is a wood for which it makes. Now, if you were
to pursue the path through that wood you would go
clean out of your way, and either get tangled up in
the rocks that overhang the sources of the Oriège,
or get down into the marshy sources of the Tet.
Neither of these districts are what you want. When
you get to the edge of the wood, which, as I say,
is about 500 yards from the point where the path

crosses the stream, you must turn sharp to your left
and go due east up a little water-course, which here
runs down beside the trees.   As you do this facing
due east, and looking up this water-course you will
see before you a ridge like any other of the Pyrenees,
with peaks upon it.   This ridge is the watershed
between the County of Foix and the Roussillon, and
is to-day the frontier of the department of the Eastern
Pyrenees, which is the modern representative of that
ancient province.   The ridge is plain enough but to
cross it is not so simple a task as it looks.   You
must not attempt to go across it by the depression
which lies immediately before you between two Peaks.
It *can* be done, but the chances are you will lose your
way in the great forest upon the further side.   The
right way is to go on due eastward up the stream
until you are right under the ridge, from which point
you must bear to your left up the bank which encloses
the gully upon that northern side.   You will notice
two peaks of rock at the point where this bank
branches from the main ridge.   You must so bear
up that you leave them both to your right, and
turning round the base of that one which lies furthest
west of the two, you will see (when you are round
the base and over the bank) a saddle just east of
you and about 600 or 700 feet below the rocky peaks
in question.   This is the *Porteille*, you will go across
it, come into the dense wood on the other side, and
there the path follows running water all the way
throughout what soon becomes a profound gorge,
until you reach open country and a few small
buildings 3 miles further down ; though the open

country, it is true, is only a small stretch of meadow between the wood and the river (a stream called the Galbe). The way is clear between the wood and stream for 2 miles more to the hamlet of Espousouille. There you must leave your path and take one which branches straight off to the right, goes down to the stream, crosses it, rises through the wood beyond, and in less than a mile from Espousouille, brings you into the considerable village of Formigueres.

I have already said that you would not easily manage this crossing in a day, even in fine weather. The Porteille is over 7000 feet high, and you may quite possibly lose your way for an hour or two in the difficult bit, but luckily there is no difficulty about camping. There is good camping ground with wood and water in every part of the journey, except the last mile of the steep going over the ridge. And you have only to choose where you will pass the night.

This is the shortest cut by far from the County of Foix into the Roussillon. If you are going down into the Cerdagne a great national road takes you from Formigueres to Mont Louis, and the distance is about 9 miles, but if you are going down into the valley of the Tet in order to climb in the Canigou you must make for Olette, for that cuts off a corner. Olette is just under 10 miles in a straight line from Formigueres, but the county road which joins them has to cross a pass and is full of windings, so that the whole distance, even if you take short cuts to cut off the long turns, is more like 14 miles. The pass, which is nearer 6000 than 7000 feet high, is 1200 feet above Formigueres, and stands just opposite that town in full view, the summit of it about 2 miles away to the south-east, but there is no need to describe the road, as it is an ordinary carriageway from the one place to the other. At Olette you are on the Tet, about 5 miles from the rail head at Villefranche. They are, however, at this moment prolonging the railway up the valley, and I do not know if they have recently opened a station west of that town.

sûr-mer

le Boulau

eret

R.

Amelie
les-Bains

rles

Scale of English Miles

Heights in Feet...... 9135.
Railways...........
Main Roads........
Other do..........
Tracks...........
Political Boundary

C. de Girere
5702

Corsady

e Tech

manya

s

Val de Valmanya

inca

Taurinya

Mt Canigou
9135
P of 13 Winds 7598
C.Portaillet

Cambret

TeR.

RADES

Fillols

Vernet

HOTEL

Br

P. de R

88

Guillem

Pl

Mantet

VIII. The Canigou

The Canigou, whichever way one looks at it, is a separate district and must be separately approached and separately travelled in. It stands apart from the rest of the range, it has a different character, and travel in it is of a different sort from other Pyrenean travel. It is not only physically cut off from the rest of the Pyrenees, indeed, its physical isolation has been a good deal exaggerated by people who have looked up to it from the plain and have not carefully noted its plan ; it is rather morally cut off by the way in which it dominates one particular province and one famous plain to the exclusion of every other peak ; so that when you are going through the Rousillon, especially along the sea coast, the only thing you can think of is the Canigou, which seems to be as much the lonely spirit of the district, as Etna does of the sea east of Sicily, or as Vesuvius does of the Bay of Naples. It will perhaps sound surprising or unlikely to those of my readers who know the Pyrenees, when I say that the Canigou is not physically isolated from the chain, it is indeed less isolated in its way than is the Pic du Midi de Bigorre, or even the Pic du Midi d'Ossau, for it is connected with the south by a high ridge which one can hardly ever see at full length from the plain, and which is, I think, only clearly observable from the frontier heights south of Arles upon the Tech. How thorough is the connection, however, what follows will show.

The Canigou is somewhat over 9000 feet in height, to be accurate 9135, yet it is but the terminal point *and not the highest point* in a long ridge which runs south westward to the frontier at the Roque Couloum. It next forms that frontier for 15 or 20 miles, and is then continued past the Port de Col Toses into Spain, where it forms the magnificent wall of the Sierra del Cadi.

A man without heart or vision would see in the Canigou nothing but the last northern point of that long range, but the political accident which makes the Roussillon French, the cross chain which springs from the Pic de Couloun and runs to the Mediterranean, and above all the aspect of the mountains from the civilised wealthy plain to the north and east (where the connecting ridge cannot be seen), and its false appearance of isolation when one observes it from the sea, all make of the Canigou, one of the most individual mountains in Europe.

There are, as I have said, many heights in its own ridge, further to the south and west, which surpass it. The Donyais is within a few feet of it, the Enfer or Gous and the Pic du Géant next door, above the valley of the Tet, are higher ; the Puigmal

just on the watershed is much higher. The summit
of the Canigou is but 1500 or 1600 feet above the
crest of the ridge in its own immediate neighbourhood, and even the lowest point in that ridge (the
Col de Boucacers) is not 2000 feet below it. Nevertheless, it produces, as I have said, an effect of unity
and of isolation, and there is not only the illusion
of its outline as seen from the north and east, but
also the fact that the mountain spreads out in a fan
of ridges from its summit to the lowlands all around,
and stands upon a broad expanded base, more or
less circular in shape, spreading from the Tech upon
the south to the Tet upon the east, north, and west.

The Canigou is not a mountain that gives one
any climbing to speak of, or that affords any
problems or difficulties. There is even, nowadays,
a carriage road most of the way up on the northern
side, but it is the best place for camping and
changing camp that you can find anywhere. All
the flanks of it are covered with a series of dense
woods ; they form a belt 2 or 3 miles deep (in places
nearly 5) and running almost continuously round
the whole mountain, a circuit of at least 30 miles.
Your choice for halting and camping places in these
woods is infinite, there is water everywhere and you
are nowhere too far from provisions. If you will
take the road from Villefranche up to Vernet you
will, at that village, be near the steepest side of the
mountain and a wood which everywhere affords
excellent camping ground. By following up the
path to Casteil and taking the track which leads
south and east from that hamlet, you are at the

inhabited point nearest to its summit, and you have wood and water up to the last mile in distance, or the last 2000 feet in height; but remember, if you wish to make for the summit by this trail, that you must always bear to the right as you walk, choosing always the right hand trail when there is a diversion, and coming out on the south side of that ridge which has the summit at one end and the Peak de Quazemi at the other. On the open part of this steep bit there is a definitely marked path which follows the left bank of the stream until it is right under the last rocks of the Canigou and then makes straight up by zigzags. If you would go the easier way which everybody takes, you must start from Prades, which is the town of the mountain, and in which anyone will show you the house where the local agent of the French Alpine Club is ready with information.

Your road goes through Taurinya (or if you start from Villefranche, through Fillols), and the new carriage road runs up the ridge between the two valleys—the valley of the Fillols and the valley of Taurinya—first over open country, then through wood until you come to quite the upper part of the Taurinya, where the road turns round the steep corner overhanging the sources of the torrent. This particular wood is called the wood of Balatg, a word that is not so hard to pronounce in Catalan as in French, for the Catalans add an " e " at the end of it.

The road does not go to the actual summit, but comes out on to the shoulder of the mountains, an open space looking to the north, north-west and east, where stands the hotel which has been put up by the French Alpine Club. This hotel is not quite 2000 feet below the highest summit which lies exactly to the south of it. The other summit to the north-east, the ridge of which comes round behind the hotel, is the Pic Puigdarbet. You must allow five or six hours to get to the hotel without haste from the valley of the Tet, and the road is somewhat shorter if you start from Villefranche, than if you start from Prades, but of the two ways, much the more interesting for a man on foot is the old way by Casteil and the Brook Cady which I first described. Here you can camp half way up the mountain without fear of disturbance from travellers, choosing, for preference, the end of the wood just under the summit, and so make that summit at dawn.

Unless you are in a hurry to get on to Perpignan one of the best ways of treating the Canigou is to go across it from the valley of the Tet into the valley of the Tech, and from Arles on the Tech to take the railway through Ceret and Elne to Perpignan.

It is of course a long way round, but it shows you both sides of the mountain.

You can hardly get right across the main ridge from the hotel; but you can take the path that goes round the northern flank of the mountain, that is, through the wood that clothes the buttresses of the Pic Bargebit, and that comes out in the valley of the Dalmanya, a torrent runing down north-eastwards from the summit.   If you are afraid of losing your way you can go down into the village of Dalmanya and up thence by a clear path from the church of the village to the iron mines under the Col de Cirere ; from that col there is a very winding highroad (of which of course you can cut off most of the turnings) which gets you down to Corsady and so to Arles. On the southern side of the mountain you can go down the path which follows the Brook of Cady, and do your best to note the peak of the Thirteen Winds which is the peak precisely due south of the main summit and 3000 feet from it at the end of the long ridge.   When you have made quite certain which is the peak of the Thirteen Winds, cross the brook, and work up if you can to the saddle immediately south-west of it, and between it and the Pic de Routat, which is a trifle lower and rises a thousand yards to the south-west of the peak of the Thirteen Winds.

This col is called the Portaillet, and the valley on the further side is called "The Old or Abandoned Pass."   When you have got across you will know why.   A wood covers its lower part, and a little brook called the Cambret runs through it, but there is no regular path, and it is a business to find the first huts, which are at an open space upon the stream between it and the wood, and quite 4000 feet below the col.

The descent is exceedingly steep, and there I leave it.

From these huts (which are called St Duillem) is a good plain path down to the Tech, and to the

little hamlet which has the same name as the river
(Le Tech) whence the national highroad takes one
in 6 miles to Arles, the more usual crossing (which
is not really a crossing of the mountains at all, but
a crossing of the ridge to the south of it) is by the
Pla de Guillem, so called because it does not go
near Guillem, and this way is as plain as a pike
staff. You take the road from Villefranche to
Fuilla, which is not quite 3 miles off, first up the
Tet, then to the left southwards up a lateral valley,
you follow that lateral valley and the high road up
it from Fuilla to Py, rather more than 5 miles on,
and southward all the way from Py a path goes
south-west up the right bank of a torrent which
comes in there. The track is quite clear and carries
you up to the sources of the stream, and to the
saddle in the final ridge which is called the Pla de
Guillem. It is a steep climb of nearly 4000 in
rather more than 4 miles. Py at the junction of
the streams is just over 3200 feet above the sea.
The pass is about 7000.

On the further side, also the track is quite plain,
pointing down due south-west through a little wood
and then over the open country. It takes you
down to Prats de Mollo, a jolly little town, the
last on the great national road and the highest in
the Tech valley. Above it the national road
becomes the local road leading to the baths and
waters.

So late as the Revolutionary Wars Mollo was of
importance and may be again, for the Spanish
armies could come over (but not with guns) from

the other Mollo, which lies beyond the frontier 7 or 8 miles off south-east, over the Col of Arras. Mollo is a little lower than Py, but the descent upon it is far less steep than was the ascent upon Py. From Mollo it is somewhat more than 10 miles to Arles by the national road down the valley.

The Canigou is so particular a thing that if a man has but little time before him, or if he already knows the other Pyrenees—he might do worse than go to Perpignan and spend a week upon that mountain. It should be remembered that you have a better chance of fine weather there than in any other part of the Pyrenees, and you will usually have dryer days upon the Tech side than upon the Tet side.

With these eight divisions I have roughly covered the chain of the Pyrenees for those who may, like myself, think that all travel on these mountains should be on foot. It is, of course, but a very rough and general survey, but it would give one, all taken together, a comprehensive knowledge of the chain. My limits have necessarily excluded very many valleys, some of which are unknown to me, such as the valley of Isaba. Among those which I have not dealt with should be considered especially the Ribagorza, which is the boundary between the Aragonese and the Catalan tongues, and runs parallel to Pallars or the valley of Esterri, and can be reached from the valley with some difficulty by Espot and the high Portaron above it, or much more easily from Viella in the Val d'Aran,

by the high Port de Viella, which leads straight
into the Ribagorza and down to Bono. There
are also entrances in and out of Andorra, of which
I did not speak, notably the Porte Blanche, which
you make from Porta in the Val Carol, a mile or
two south of Porté. This way involves two cols,
one very high one, the Porte Blanche, another
lower one immediately after, the Port de Vallcivera.
It is, however, the shortest way from a French high
road to Andorra, the Old. There is another way in
and out of Andorra, very little used, by the Col de
la Boella from Ordino to the Val Farrera. All the
Basque valleys, besides those I mention, and notably
that of the Isaba, are places that should be known,
and of the passages over the range, which I have
not dealt with in detail, one, the road from St Girons
to Esterri by the Port de Sąlau, will soon be an
International highway. It presents no difficulties
and no very considerable interest. But if the
traveller finds himself by some accident in St
Girons with but a day or two in which to see
Spain, here is a very easy way of getting over
into what is still one of the remotest parts of that
country.

VII

INNS OF THE PYRENEES

THERE is nothing more necessary to the knowledge of a district if one desires to enjoy travel in it, than to have some directions upon its inns. I cannot pretend in what follows to give any complete list of the inns which the traveller will find in the Pyrenees, but I will try to do what the guide books do not do, and that is to indicate what an Englishman, especially one on foot, may expect in the different valleys. The foreign guide books rarely do this well: the Scotch and English guide books never: for the general phrases which they use about inns and hotels leave one as full of doubt and terror as though nothing had been said about them, and they always fail to speak good or evil of the *people*, the *cooking*, and *the wine*—which are the three main things one wants to hear about.

First then, as to the difference between the Spanish and the French side.

Though the Basques are one race upon either side of the frontier, and the Catalans also, yet a single rule governs the whole length of the chain, which is that French cooking and French hours are to be found to the north of the political frontier, and Spanish to the south. This is a matter in which the difference of Government has, in the course of some generations of travel, produced a very marked effect. The Val d'Aran, for instance, is geographically and racially French. Its river is the Upper Garonne, there is no obstacle between it and the French plain, but only one good descending road to unite them both; yet your experiences of an inn in the Val d'Aran will in general resemble your experiences of an inn beyond the mountains in the purely Spanish valley of the Noguera.

Similarly the neighbourhood of Saillagouse and all the French Cerdagne is geographically and racially Spanish, the river running through it is the Upper Segre (a tributary of the Ebro, and one road with no obstacle at the frontier, unites the French to the Spanish portion of the valley, yet the hours, habits, cooking, and everything in the inns of the French Cerdagne are French, in those of the Spanish Cerdagne, Spanish; and generally you must be prepared, when you cross the frontier, for a different kind of hospitality.

The French rule of an inn is probably well known to all who will read this. The coffee in the morning, the first meal at or a little before mid-day, the second at six or seven at the latest, and so forth. In Spain they will give you chocolate for your first meal. Your mid-day meal will be at the same hour as the French, but your last meal much later : eight is a usual hour. In France, if you ask for food at an odd time it will be prepared for you ; in Spain also but only with incredible delays, and you find universally upon the southern side of the frontier, this difference from the French that the table d'hote or common meal is prepared only for a fixed number of guests. Newcomers, even if they reach the place two hours before the hour of the supper, have it separately cooked for them, and will suffer a corresponding delay. Here is a national custom which nothing can change, and which is as old as the hills. It was even once universally the habit to have a separate little cooking pot for every guest, and in certain inns that habit is still continued. It is in the last degree inconvenient, and when one has pushed on to the end of some very long day, to shelter and food, it is exasperating. One sees the local people who have done nothing, eat a hearty meal ; and one waits an hour or two hours before one is served with a crust. But you can no more change it than you can change any other national habit, and you must be prepared for it on the Spanish side wherever you go. All the details of the cooking are different too ; notably these : that for some reason or other, the Spaniard is careless of his oil, or perhaps prefers oil to have a taste of carelessness about it : in places of rancidity. His wine is quite different from the wine of the French. It comes up

to him from the hard plains of the Ebro ; it has
been kept in wine skins and tastes of them. As a
rule drink water with, or better still after, Spanish
wine The French wine in these hills (save in the
Roussillon) comes from the plains of the Garonne,
and has been kept in wood. It has the taste with
which we are familiar in this country ; the Spanish
wine has a roughness, a strength, and a memory of
goat's skin, with which, until he comes to Spain,
no northern man can have any acquaintance at all.

It must not be imagined that Spanish accommodation is cheaper than French ; comfort for comfort,
it is, if anything, a little dearer. But the Pyrenees
are cheap everywhere, save in one or two watering
places. Nearly every inn upon either side, however small, can furnish you with a guide, but not
every inn with mules, and still less can you depend
upon a horse or a carriage, even in places which
stand upon the few great highways. If you must
hire mules, you will always be able to find one in
the village where the inn stands, but, for some
reason connected with their local economics, the
people of the inn are sometimes actively opposed
and often indifferent to your hiring one, and if they
tell you that there is no mule to be had (which is
their way of opposing you) you must then saunter
out and bargain for one with some rival, but remember that you can always get one : all these
mountains are covered with herds and droves of
mules. Yet mules are expensive, from 300 to 500
francs to buy, or even more ; from 10 to 15 francs
per day to hire, with the man who accompanies you.

Remember also, if you have a choice where to hire, that they are better by far upon the Spanish than upon the French side. As for horses and carriages, I will, when I speak of particular inns, mention the few places where I know they can be hired.

A further difference between the French and Spanish side is that, on the whole, an inn upon the Spanish side is less likely to be clean. This does not mean that they are generally uncleanly, very far from it; the houses of the whole of the Basque country on either side are excellently kept, and this is generally true of Catalonia also, but the little hamlets, in the highest valleys which are doubtful upon both sides, are usually worse upon the southern. In every case, of course, you must 'ask the price of rooms, they expect it, and it is best to ask the price of meals as well. If you do not bargain in this manner, they think of you as of someone who is deliberately throwing money away and they very naturally hasten to pick it up. I remember one meal in the very unsatisfactory town or village of Llavorsi, which was as unsatisfactory as the place itself, and for which a violent Catalonian woman would have charged us the prices of Paris because we did not bargain beforehand, and this, note you, in a place where no one ever comes, which is on the road to nowhere, and which does not see tourists perhaps, or even travellers, once in six months.

In every valley there is some one inn which, if you are wise, you will choose, and which it is worth one's while modifying one's plans to visit. I will set down those which I know, beginning as I have done throughout this book, at the western end of the chain, and following it to the east.

In the Baztan, a Basque word for tail, for the valley resembles in shape the tail of a rat, though the other *Bas*tans in the Pyrenees, out of the Basque countries, derive their name from the Arabic word for garden, Elizondo should be your halting-place. Here there are two hotels, one old and one new, the old one in the very middle of the town on the high road, the new one a little to the north, just off the high road.   This new hotel is kept by one Jaregui, and in the chief feature of all good hotels (I mean the courtesy and zeal of the management) it is far the best, not only in Elizondo, but in the whole valley.   If you should wander on to Pamplona, I can give no advice, but it is a large town where a man may have pretty well what he wants according to the price he pays.   My own experience of it is of lodging in small eating-houses, not in a regular hotel, but I understand that the Perla and the Europa are the two best hotels, and of these two, people, as one travels, single out the Europa.   On the road from Pamplona to Roncesvalles, there is no good stopping-place.   At Erro, as I have said above, there is but one inn and that a very bad one. Burguete is, however, a very pleasant village, and the Hotel des Postes is praised by those who have stopped there.   Unless one is caught by night, or in some other way impeded, it is unwise to eat or to sleep at Val Carlos, the contrast between French and Spanish methods is nowhere more violent

20

whether in the matter of cooking, or of delay, or of wine, or of any other thing, than at this corner of the frontier; but it is to be remembered that if you need a horse and carriage you can always have it at Val Carlos for going on into France, and at St Jean Pied-de-Port you are in the best halting-place for the valley of the Nive and the whole Labourd, just as Elizondo is the best halting-place for the Baztan. St Jean Pied-de-Port is large enough and frequented enough to have some choice of hotels. You had much better go to the best, which is the Central. The reason it will be worth your while to do this is, that though it is the best hotel in a town to which many rich people come, it is as cheap as it is good. It will always have a carriage for you if you want it, it has a garage, and it is the best centre from which to start upon any of the roads around; and if you should be coming from the north and going south there is a public service from this hotel through the pass as far as Pamplona.

In the next valley, that of the Soule (the river of which is the Saison, and the chief town Maulèon) let Tardets be your headquarters. It has one of the most delightful inns in all the mountains, remarkable among other things for having various names, like a Greek goddess. Sometimes it is called the "Voyageurs," sometimes the "Hotel des Pyrenees," and it is entered under the arcade of the north-west corner of the Market Square. There you may dine in a sort of glass room or terrace overlooking the river, and everyone will treat you

well. It is, I say, one of those places that would
make one hesitate to go on further into the hills
the same day, but if one does, one will find the
unique inn at St Engrace, which I have already
mentioned, one of the best that the smaller villages
have; it must always be remembered, of course,
that these upland hamlets give one nothing but
their own fare, and usually a bedroom that is
reached through some other, but the beds here
are good and the cooking plain. This is the first
house in the village on the right as you come in,
and as in Elizondo, Jarégui is the name. Remember
that they have various sorts of wine, and ask for
their best, for even their best is but 6d. a bottle,
and their worst is not so good. In the valley
between Tardets and St Engrace, before you leave
the main road, you pass by the hotel of Licq,
"Hotel des Tourists." Licq itself you leave to
the right beyond the river, but this hotel is built
upon the highroad. Here is a good place for one
meal, though there is no point in sleeping there,
yet if one is caught by some accident, one will find
it comfortable enough; a little bothersome in press-
ing one to take guides.

The next valley, the Val d'Aspe, and its prolonga-
tion on the Spanish side, the Val d'Aragon, contain
many inns, the more important of which should be
known before one approaches them.

In Oloron itself, there are two good hotels of
which the Voyageurs is perhaps the best, and
there is, of course, every opportunity, in such a
town, of hiring horses and carriages. There is

also, it must be remembered, a public service twice a day up the pass as far as Urdos, and a very cheap one; just over three francs for the whole 20 miles. It will be possible also at Oloron to hire a pair of horses and a carriage if one wants one for several days to go into Spain and back by way of the Val d'Ossau.

There is no occasion to stop, whatever be your mode of travel, between Oloron and Bèdous, but should you take up your headquarters at Bèdous (which, it will be remembered, is in the midst of the enclosed plain which characterises this valley), make the Hotel de la Paix your headquarters. You will be best treated there, and it is the best centre for information upon the surrounding mountains. Accous is slightly larger than Bèdous, but it is off the road and therefore less used to travellers; also it is less comfortable. So if you stop in this plain at all, stop at Bèdous.

Your next point will be Urdos, there is nothing of consequence between.

Urdos, having been, for so many centuries between Roman civilisation and our own, the end of the proper road over this chief pass and the jumping-off place for the mule tracks and for Spain, has many inns for its size—(it is no more than a hamlet)—but of these I will unhesitatingly recommend the *Voyageurs*, which is one of the last houses on the left of the village, having at the south end of it over the road a jolly little terrace where one dines. The drawback of Urdos is that one *may* get bitten, and speaking of this the sovereign remedy is camphor, or rather

I should say, the sovereign preventive, for all animals that bite hate the smell of camphor. But for that little drawback, Urdos is delightful and nothing is pleasanter in Urdos than the Hotel de Voyageurs, also if you go to this hotel you are following the line of least resistance, for it is in some mysterious way related to the man who drives the coach. Remember that Urdos is accustomed to every form of halt, and though it is difficult to buy things there, there is a barn for motors—and also, I believe, relays of horses for carriages.

Your next village on this main international road is Canfranc in Spain. It is just over 14 miles off with nothing but a refuge and the pass of the Somport between. The hotel is the Hotel Sisas, from which the public coach starts for Jaca (costing three pesetas); the cooking is doubtful, the wine so-so, and the people are a little spoilt, but they are very ready with horses and used to hiring them, and you can always hire a carriage or get a relay for Jaca, which is 16 miles further down by a road with no steep hills, and for the most part nearly flat. At Jaca the hotel (which I have already spoken of) is the Hotel Mur ; it is excellent in every way, clean, cheerful, and not too simple in its customs, with various wines, and a knowledge of more than the Castilian tongue. The mention of this leads me to add to what I said above that the language stops very suddenly at this central frontier, or at least south of it. There will be people who will understand Spanish almost anywhere in Bearn because the local dialects are Spanish in character, but the common French of Paris means nothing to the people of Aragon and Sobrarbe ; you may be in quite a big place and find no one for a long time who will understand you, while in the small hotels and inns right up against the frontier, they do not follow a word of the language.

Of the inns of Biescas I cannot speak from experience, nor of those of Panticosa, though they say that the only useful one in Biescas is the Hotel Chauces, while Panticosa has any number of places with such names as "Continental" and "Grand," and masses of lodgings as well, among which I imagine the only choice is to take the best ; nothing is really dear there, except in the month between the middle of July and the middle of August. Of Sallent, however, I can speak. There is but one inn in the place ; it has many names but is best known by the name of the man who owns it, and his name is Bergua. It is an astonishing mixture. The owner is wealthy and good-natured, but you do not hear the truth about things for it is coloured by self-interest. The place is clean, but slow even beyond the ordinary of a Spanish inn. The cooking is neither one thing nor another, the wine is not bad. It is a place where you may spend one night, but not two. You will leave it without enthusiasm, but without regret.

Next, following the itineraries I have given, comes Gabas, and here is as pleasant an inn as you will find in the whole world, it is called the Hotel des Pyrenees, and of the several hotels it is the dearest. The family of Baylou keep it and have inherited this soil for generations. It is an ancestor of theirs that planted the delightful Mail outside and set up the charming little fountain there. They are used in this house to every sort of gentlemanly habit, they pay no attention to the clothes in which one comes, and they understand all those who love to wander in the hills. Everything is clean and good about the place, they will give one well-cooked food in many courses at any hour. There is but one criticism to make and that is in the matter of horses and carriages ; these are dear, and the good and the bad cost the same money, for there is here a monopoly of the valley, and if you do not take their vehicle, you must walk to the rail-head, 8 miles lower down. Also if for some reason you must drive or get a relay of horses, the longer notice you give the better, for there are few animals to be had.

Further down the valley is Eaux Chaudes, a dreary place, incredible from the fact that it was here that much of the Heptameron was written! If a man must stop there, let him ; of the sad gloomy barracks, take the largest and the dearest, which is the Hotel de France. Laruns, at the foot of the valley, where again you are unlikely to stop, but where you may be caught, has the Hotel des Touristes, where also horses and a carriage may be hired, and whence the omnibus goes to Eaux Chaudes and to Eaux Bonnes. This last place, like Panticosa, is a place one can make no choice in, it is crowded with the rich, and where the rich have spoilt things, the only rule I know is to plunge and take the dearest—which is the Hotel des Princes—if you will not do that you must choose for yourself.

The next valley, that of the Gave de Pau, has in it four towns, Lourdes, Argelès, Cauterets, and Luz. Lourdes, like all cosmopolitan towns, is detestable in its accommodation, and to make it the more detestable there is that admixture of the supernatural which is invariably accompanied by detestable earthly adjuncts. Were it not so the world would be perfect: but it is so, and honestly one cannot say that any one hotel at Lourdes is better than another, only here again if one is compelled to stop for a night, one cannot do better than the best which is nominally the Angleterre. Avoid the hotels that have Holy names to them, they are usually frauds. If you go to Lourdes as a pilgrim, prefer the religious houses (which take in travellers). If the Angleterre is too dear for you, the Hotel de Toulouse is not to be despised; it will take you in at 10 or 12 francs a day. Argelès, up the valley, is a very different place, it is a little hurt by the neighbour-hood of Lourdes, and by the stream of travellers who pour up and down its main road to Cauterets and to the sights of Gavarnie. Nevertheless it remains a French country town, and the fairly dignified capital of a district. The Hotel de France is excellent and, by the way (a thing always to be mentioned when one is speaking of hotels in the Pyrenees), it is ready at any time to furnish horses, and has, of course, a garage. At Luz stand two hotels facing each other on either side of the road, I cannot remember the names, or rather I cannot remember which is which, but anyhow take the one on the right of the road as you look up the valley, or as you come up from the station, that is, the one upon the western side. They are polite, and that makes all the difference in one's relations with people whom one does not often meet.

Gavarnie, overrun as it is (and it is hideously overrun), has a very tolerable hotel, clean, and not too dear. The reason is that the people who come to the place usually go away on the same day, and that therefore there is some anxiety to please those who stop. Another inn, up under the mountain, is not so much to be recommended. Of Cauterets everything can be said—and much more—that was said of Eaux Bonnes, you are at the mercy of a place which the rich choose to have ruined, and apart from their vulgarity you will have that noise which accompanies them in all their doings, this sort of place in the Pyrenees is luckily not common, and when it is tolerable is tolerable in proportion as it is national. Cauterets is almost as international as Lourdes, and for anyone using the Pyrenees as I use them in this book, it would be madness to stop there. Bagneres-de-Bigorre is better, though it is something in the same line. It is better because it has something of a past and a history, and is, like Argelès, the chief town of its district. The Hotel de Paris is the best, but it is very expensive, and I believe, though I do not know, that the Hotel des Vignes in the Rue de Tarbes is good

among the moderate places.   But the rule holds
here, as everywhere, that where rich people,
especially cosmopolitans, colonials, nomads, and the
rest, come into a little place, they destroy most
things except the things that they themselves
desire.   And the things that they themselves desire
are execrable to the rest of mankind.

Arreau, in the next valley, merits a more particular
attention.   It is thoroughly French, and here you
will find side by side with the expensive places (for
even Arreau has its Hotel d'Angleterre which,
however, to tell the truth, is not ruinous) a most
delightful little place called the Hotel du Midi,
where sensible people go.   I am speaking on the
testimony of others, but on good testimony.   It is
a place smelt out by the infallible nose of the French
professional class.   It has a garage, and will tell
you where to get carriages, though I believe it has
nothing but an omnibus of its own.   Luncheon is
2s., dinner 3s., and when you are paying more than
that anywhere in the Pyrenees you are being
cheated.   But for some odd reason this excellent
house charges you extra for your coffee.

Right high up this valley is Vielle where there is
one hotel, the Hotel Mendielle, this is the one you
must ask for if you find yourself caught here, and it
is just the place at which one might be caught if
one got into the wrong valley from a col in the
Sobrarbe, or, if, in coming up the Gave, one had
not made way enough by night; I know nothing
for or against this hotel, and I believe it to be the
only one.   The little village of Aragnouet, which is

at the very end of the road under the last precipices, has an inn of the quality of which I know nothing.

The next valley is that of Bagneres-de-Luchon. Now it might be imagined, seeing what rich places are in the way of hotels, that Bagneres-de-Luchon (being by far the richest place in the Pyrenees) would be hopelessly the worst, and that, as nothing good could be said about Cauterets, and as there was precious little choice in Eaux Bonnes, Luchon would be a place to despair of in the matter of hotels, but on the contrary it is a place to discuss.

Even if Luchon were as detestable as the Riviera, one would have to come to it because it is the knot and reservoir of all mountain travel. The valley strikes so deep into the hills, brings the railway so near their summits, and is so exactly situated at the "fault" spoken of so frequently in this book (the break in the Pyrenean line where the landscapes and peoples of the chain meet) that it is difficult not to pass through Luchon at one time or another during any length of days past in these hills. Even if you make a vow to clear Luchon, you may find yourself caught in any one of twenty surrounding barbarisms with a bad foot or no money, and compelled to set a course for this harbour. Moreover Luchon is by no means the vulgar place its riches ought to make it. The fashion for it was first made by reasonable people, many Spaniards come and help to give the place its tone, and perhaps the very extremity of evil corrects itself, and Luchon, being so crammed with wealthy people, knows its own vices better than places just a little less rich, and it is therefore more tolerable. At any rate the problem of sleeping at Luchon is easily solved in July and August because all prices are pretty much the same, and you cannot depend upon the printed prices at all. For pension it is otherwise. There are fixed prices and they are not exorbitant for such a place. A very clean, decent, rich hotel is the hotel d'Angleterre, where, if you stop some days, they will charge you, I believe, about 12 francs a day. There is a place for poorer people called the Hotel de l'Europe; all its prices are cheaper, but it has this drawback that you get nothing national. It is clean and there is a roof over your head, but you get neither French comfort nor French discomfort, and you are paying a little less for things a great deal worse, notably in the matter of food. The bold who fear nothing will go and stop at the little village inn called the Golden Lion, which is near the old church and existed before wealthy Luchon was born or thought of. Here the bold will consort with Muleteers and the populace in some discomfort. One of the best uses to which one can put Luchon is to eat in it, and for sleeping to go outside and camp in the woods : and the best place for the passer-by to eat is the Café Arnative on the main street ; its cooking is very good indeed, and the wine really remarkable ; it is such good wine that one wonders why they give it away, and every year as one returns to the place one fears it may have ceased, but it continues. Speak to the manager in English for he knows and loves that tongue, or in Spanish or in French. In

the use of the hotels and restaurants of Luchon,
however—always excepting the Golden Lion—
remember that they are snobbish about clothes,
and that even two days in the hills puts you well
below the standard which they can tolerate. I
confess that when I have had to use Luchon, I
have depended upon clothes which were waiting for
me at the station; and it is not difficult to use
Luchon as a sort of half-way house in this matter,
leading the right life in the western mountains,
coming down to Luchon to find one's luggage, dress-
ing up, plunging into worldly pleasure at Luchon,
sending one's luggage off again to Ax or Perpignan,
and then taking to the eastern hills for another bout
of poverty.

In the Val d'Aran, next to the valley of Luchon.
there is but one place where one is likely to stay,
and that is in the town of Viella, which is the
capital; for the Val d'Aran is a small place, and
there is no advantage in stopping anywhere else,
The Posada Deo is that which I know best and is
good but of course Spanish; the cooking is a sort
of mixture of Spanish and French, but the time you
have to wait for it and in the manner in which it is
given you is wholly Spanish. The wine also (oddly
enough!) is Spanish. It ought, on the Garonne, to
be of the Garonne, but the customs interfere.

The Catalan valley, south and east of the Val
d'Aran, the valley of Esterri, has, in that town, a
good little hotel, the Hotel Pepe. The people are
thoroughly Catalan in their love of money and
therefore you must bargain. Whatever you do, do

not stop at any of the other places in the valley, it is even better to go through a storm than to risk Llavorsi, or worse still Escalo, but on the far side of the hill and of the port called St John of the Elms there is a most delicious inn, with an old innkeeper of the very best, at Castelbo.

To return to the French side ; if you go by train to St Girons you may likely enough change at Boussens, the station has not (or had not) any buffet, but there was (and I hope is) a hotel opposite it where people travelling by train ate ; the cooking here is the best in the whole of the Pyrenees, which is saying a good deal. At St Girons itself there is not only good cooking, but the wine which Arthur Young admired, and which was well worthy of his admiration. Do not go to the best hotel (which is the hotel of the Princes and of the Alpine Club), but to the next cheapest which is called the Hotel de France ; at least I have found this last to be excellent and cheaper for its quality of food and drink and repose than any other in all this chain. These things change quickly, what was true so short a time ago may not be true now ; but so, at least, I found it.

In the valley of the Ariège it is always well to make Ax your sleeping place, for Ax, though there are waters and though the baths make the prosperity of the place, is a very pleasant little town and the right beginning for the mountains, whether you are going by the main road into the Roussillon, or up the Ariège in the Carlitte group, or again over the main range into Andorra. At Ax there are two rival hotels, the Hotel de France, and the Hotel Sicre. The latter is a little cheaper though both are cheap, and while I know the second one best I should recommend the first ; it will take you in at less than eight francs a day, and seems the more carefully kept ; both have garages. The Hotel Sicre suffers somewhat from being directly attached to its Thermal Baths. If you are going to explore the wild country of the Upper Aston, you must start from Cabanes lower down on the railway. There is no need to sleep there. The valley above it has some of the best camping places in the Pyrenees. But it is worth knowing the name of the hotel, which is "Du Midi." The whole place is, of course, quite small and cheap.

On the high road into Rousillon choose Porté, primitive as it is, and avoid *Hospitalet* (on the hither side of the pass of Puymorens) like the plague. Hospitalet and the village just before it, Merens, are for some reason or other quite spoilt ; I fancy tourists come up so far as these two without going over the pass which they find too much trouble, and that their coming and going has spoilt the two places : at any rate they are detestable. They overcharge you and treat you with contempt at the same time.

Porté, though it is but a few miles further on, is quite different. Here is one rude inn, as cheap as the grace of God, and kept by the most honest people in the world ; Michet by name. It is thoroughly Spanish in character (for remember that Porté, though politically in France, is on the Spanish side of the main range, and that the pass just above is on the watershed); the animals live on the ground floor, the human beings just above them. You will never regret to have slept at Porté.

As you go on into the plain of the Cerdagne you will find a good inn at La Tour Carol: not exactly enthusiastic in their greeting of the traveller, but polite. It is quite a little place of only half a thousand inhabitants, and you cannot expect much from it, but it is better than Saillagousse where they are most unwilling.

Up the road to France from Saillagousse, at Mont Louis, is a hotel of which I can speak but little because my only experience of it was late on a holiday night when everything was very full, but it is substantial, it is cheap and I have heard it praised. It is called the Hotel de France, and it is a starting-point for the omnibus down to the rail-head at Villefranche in the valley above which rise the flanks of the Canigou.

On the Canigou itself, standing upon a platform a few hundred feet below the summit facing the Mediterranean and one of the greatest views in this world, there is now an inn which you must not despise though it does happen to be somewhat tourist. It is only open for the end of June, July, August, and September, though one can sleep there at other times of the year if one asks at Prades for the housekeeper; he comes down to that town through the winter and is known there.

In Perpignan (by the way) go to the chief hotel, for the hotels of that plain can be very vile when they try. This hotel is called "The Grand" and it stands on the quay of the smaller river just within the old fortifications. There is a delightful little restaurant in Perpignan called the Golden Lion, it is well to order what one wants some hours beforehand, and to take their own recommendation about wine. Perpignan is so twirled and knotted a town that I can give no directions for finding that Golden Lion, where it lies in its little back alley called the Rue des Cardeurs, save to tell you that it is but 200 yards from your hotel, and that the Rue des Cardeurs is the second on the *left* as you walk away from the main front of the cathedral ; or again, the *first* on the left after you have crossed the Place Gambetta. Anyhow, Perpignan is a small place and anyone will show you where this eating-house is, and it is a good one. Down the Cerdagne in Spain, at Seo de Urgel, there are two or three hotels, and one of the second class called the Posada Universal or Universal Inn which merits its name ; you will do well to stop there for it has a pleasant balcony overlooking the valley, with vines trained about it ; and the people look after you.

As to the inns of Andorra your best plan is to stop in the capital, that is, in *Andorra The Old* itself, where the Posada is called the Posada *Calounes*, and is quite a little and simple place. The entry into Andorra, however, is not always easy. If you make it from the north, mist may delay you, even on the grassy Embalire Pass, and may keep you for hours on the higher crossings of the range, even when it does not defeat you altogether. You may therefore have no choice but to stop at one of the little villages ; but it is a poor fate, for they are full of bugs and fleas and appalling cooking, though the people are kindly enough. The inn at Encamps is the only one with which I am myself acquainted among these smaller places ; there also it is vile.

I have omitted so far to speak of the inns in the Sobrarbe. That of Venasque is the largest and most used to travellers. Like all Spanish inns the life of the people is upstairs and the life of the animals below. It is clean and seems to be continually full of people, for there is quite a traffic to and from this mountain town. The inn has no name in particular that I know of, but you cannot miss it. Guide books call it " Des Touristes," but I never heard anyone in Venasque give it that name. You have but to ask for the Posada, however, and anyone will show it you. It is in the first street on the left out of the main street as you come into the town. As to the cost of it, it is neither cheap nor dear ; but (as I have said is common to the Spanish Inns), it is a little on the side of dearness. A friend of mine with three companions and two mules found himself let in for over £3 for one night's hospitality, on the other hand, I myself, some years after, with two companions, passed two nights and the day between with everything that we wanted to eat, smoke, and drink, and we came out for under £2. The mules perhaps consume.

In all Sobrarbe there are but the inns of Bielsa and Torla (I mean in all the upper valleys which I have described) that can be approached without fear, and in Bielsa, as in Venasque and in Torla, the little place has but one. At Bielsa it is near the bridge and is kept by Pedro Perlos; I have not slept in it but I believe it to be clean and good. El Plan has a Posada called the Posada of the Sun (*del Sol*), but it is not praised; nay, it is detested by those who speak from experience. The inn that stands or stood at the lower part of the Val d'Arazas is said to be good; that at Torla is not so much an inn as an old chief's house or manor called that of "Viu," for that is the name of the family that owns it. They treat travellers very well.

This is all that I know of the inns of the Pyrenees.

VIII

THE APPROACHES TO THE PYRENEES

A TRAVELLER from England, on considering his approach to the Pyrenees, must first
appreciate the road heads or starting-places
whence his travels to the Pyrenees may be made,
and it is convenient to regard that one to which
access can be had by rail. These points are eleven
in number — St Jean Pied-de-Port, Mauléon,
Oloron, Laruns, Argelès, Bagnères-de-Bigorre,
Arreau, Bagneres-de-Luchon, St Girons, Foix, and
Villefranche, which last is the highest point to which
the rail will take one from Perpignan.

One can get nearer the main range by light
railways in certain places. Thus from Mauléon a
steam tramway will take one some miles nearer the
hills, to Tardets. From Lourdes the train goes up
the valley several miles, and light railways go to
Canterets and Luz, and from Foix there is a
considerable reach of rail, as far as Ax-Les-Thermes,
all up the valley of the Ariège, from which lateral
valleys on every side enter the high mountains.
Nevertheless, if one knows how to approach these
eleven stations, and something of the hours of
arriving at them, the slight extensions in the

three cases named can easily be looked up, and there is no need to burden these pages with them.

Of these eleven, the first four, St Jean Pied-de-Port, Mauléon, Oloron, and Laruns, belong to the western section of the range, and are approached from Bordeaux. Another four, Arreau, Bagnères de Luchon, St Girons, and Foix belong to the central and eastern section of the range, and are approached by way of Toulouse, while the two intermediate ones, Lourdes (and its extension up the valley) and Bagnères de Bigorre, may, according to the convenience of trains, be approached with equal facility from either direction.

There remains Villefranche, the chief station under the Canigou, and the centre for the extreme eastern end of the range. The approach to this short and distant part of the Pyrenees is through Perpignan.

By whichever road one approaches the Pyrenees, and from whatever town at their base one proposes to make the ascent of them, one leaves Paris by the Orleans line, choosing for preference the great new station on the Quai-d'Orsay, though if one is driving across Paris with no time to spare, it is better to catch the train at the Austerlitz station a mile or two further down the line where all the expresses stop, as the departure from that station is ten minutes later than from the Quai-d'Orsay. But the Austerlitz station is old-fashioned; all the conveniences of travel are gathered at the more recent terminus, and if one has any time to spare it is always from the Quai-d'Orsay that one should start.

Arrived whether at Bordeaux or at Toulouse, one changes from the Orleans system to the Midi. This is not an absolutely accurate way of putting it, because, as a fact, the Orleans only enjoys running powers to Toulouse, along the main ϵ ϲpress line, but this is roughly the best way of putting it to make the reader understand the way in which the systems join.

With these connections, the first journey is made to Bordeaux, to Toulouse (or, in the exceptional case of the extreme east end of the Pyrenees, to Perpignan), and the journey forward from each of these towns is calculated upon another time table, and is often taken on a different train.

To reach St Jean, one goes on from Bordeaux to Bayonne and changes there. To reach Mauléon, one goes on from Bordeaux to Puyoo and changes there ; to reach Oloron or Laruns, one goes on from Bordeaux to Pau and changes there.

Roughly speaking, those who want to take the journey easily, without night travel, will find it necessary to sleep in Paris, to sleep again at Bordeaux (or somewhere further down the line, as at Bayonne or at Pau) and only on the third day to proceed to the towns from which they will begin to climb, whether that town be St Jean, Mauléon, Oloron, or Laruns. For this purpose they must take the morning train which leaves Paris in the neighbourhood of eight (though the hour is slightly changed in different years), and gets to Bordeaux somewhat before five.   It is then possible to go on the same evening to Bayonne, and, if one goes first class, to get on the same night also to Puyoo or to Pau, but in all cases arrival at the foot of the mountains will not be possible until the next morning.

Those who are content to suffer night travel will find an excellent and convenient train leaving Paris always some few minutes after ten (this year it was 10.22), reaching Bordeaux by seven, and putting them at any one of the mountain towns at, or a little after, noon.   Thus, a person leaving London upon Saturday morning, will, if he travel only by day, reach any one of the western approaches to the Pyrenees on the mid-day of Monday, but if he will consent to a journey by night, he will save exactly twenty-four hours and arrive at noon (or in the early afternoon) of Sunday.   The gain of twenty-four hours, by an apparent sacrifice of only twelve, is due to the nature of the connections between the small mountain lines and the main lines.   His return tickets, going in the cheapest manner, second class from London to the mountains and back will vary according to the mountain town chosen, from a little under £7 to close upon £8.   For myself, going third-class and taking advantage of special tickets—circular and other—I have often done it for just over £5, but I am giving ordinary second-class fares.   For first-class fares one must add nearly £3 to the above rates, which makes them vary from £10 to £11 instead of from £7 to £8.

The approach to the intermediary towns of

it by night, one must leave Paris at seven o'clock in the evening in order to get to Perpignan for lunch, or at half-past eight to get in at two. It is no way to approach the Pyrenees, unless one happens to be taking a journey down France for other purposes which will lead him towards the districts of Narbonne and Perpignan, and it must be noted that even the expense of taking the Barcelona express, which runs but twice a week, does not save one, because that splendid train is designed to pick up travellers from Lyons, Germany, and the centre of Europe, rather than from Paris.

Speaking, by the way, of the great European expresses, the Sud Express, leaving Paris just after noon, will get you to Bordeaux at seven, to Bayonne before ten at night. One thus gets the advantage of what I believe to be the most rapid service in the world and of paying for it—and of leaving Paris an hour or two later; but one does not gain anything in the full time to the mountains over the trains I have indicated.

No other approach to the Pyrenees save these by railway from the north will be of use to most travellers from England. It is, however, worth mentioning that there is a line of steam boats from Liverpool to Bordeaux twice a week, and that if one is taking the Pyrenees on one's way home from Italy or the south of France, it is a saving of every sort of worry to make straight for Toulouse from Narbonne and not to attempt the beginning of one's journey by Perpignan unless one means to confine oneself to the Cerdagne and the Canigou.

There is no really good and fast day-train to Toulouse by this cut-across, but there is a good night one.

The approaches from the south, in the rare case of a traveller who may take the Pyrenees on the way back from Spain, are all difficult with the exception of the line from Saragossa to Jaca. A main line leads of course from the capital to Saragossa, there one must cross the Ebro to the station upon the northern bank. The train to Jaca goes by Huesca and it takes all day, but it is worth doing in order to get within a day's walk of the main range.

From every other centre, except from Pamplona, the Pyrenees are hopelessly distant. Seo and the Catalan valleys depend upon Barbastro as does the valley of the Cinca in Aragon, but it is a most tedious journey in stuffy omnibuses followed by an equally tedious day and a half or two days upon a mule before you find yourself in the high Pyrenees. Pamplona is, roughly speaking, one day's walk from the heart of the mountains, and no other town, excepting Jaca, upon the railway on the Spanish side is worth considering as a rail head.

It should be noted that there is during the summer months a motor car service between Pamplona and Jaca, which goes along the valley of the Aragon and covers the distance in the better part of a day.

INDEX

A

Accous or Bedous, plain of, 219-221
Agra, river, mentioned, 25; valley of, 203
Aiguestoites, Port de, 250
Albigenses, crusade against, its meaning and results, 66
Alfonso el Batallador, 73
Alpargatas, 168-169
Alps, contrasted with Pyrenees, 33
Andorra, history and character of, 267-268
—— forms with Catalan valleys a district of Pyrenees, 259-275
—— how reached from Ariège, 259-268
—— posada of, 321
Anicle, Col d', 236-237
Anie, Pic d', its position on first axis of Pyrenees, 8
—— boundary of the Basques, 51, 213
Aphours, brook of, 207
Aragnouet, 250
Aragon, river, mentioned, 25
—— valley of, easy connection of, with valley of Gallego, 218; described, 222
—— kingdom, named after river, 25; and Béarn, their position on the range, 49
Aran, Val d', see "Val"
Arrazas, valley of, 233-234; Inn there, 323
Ariège, sources of, 265
—— valley of, position of on axis of Pyrenees, 8; forms old county of Foix, 18; in connection with that of the Tet, 283-290
Ariel, Pic d', 284
Arles, on Tech, 295-296
Arras, Col d', 298

Arreau, hotels at, 314
Arrouye, Pic d', 251
Aspe, Val d', 219
Aston, upper, adventure of author upon, 148-152
—— river, advantages of district of, 260
Ax, way from, to valley of Tet, 286-289; hotels and baths of, 319

B

Bagnères-de-Bigorre, hotels at, 313; de Luchon, see "Luchon"
Baigorry, valley of, 201, 202
Balatg, wood of, on Canigou, 294
Bambilette, port, 210
Bargebit, Pic de, on Canigou, 295
Barrosa, stream of, 240
Barroude, pass of, 250
Basque, place names found throughout Spain and Pyrenees, 51-52
—— Valleys, a district of the Pyrenees, 200-213
Basques, their position on the range, 49; Pic d'Anie, their boundary, 51
—— no Roman record of, 60-61
Bathing, dangerous when fatigued, 197
Batallador, surname of Alfonso, 73
Bayonne, road from, to Pamplona described, 131-135
Béarn and Aragon, their position on the range, 49
Béarn, Roman name of, 60; with Navarre and Roussillon, last exceptions to French sovereignty north of Pyrenees, 66
Bedous, Hotel de la Juste at, 221

CPSIA information can be obtained at www.ICGtesting.com
Printed in the USA
LVOW04s1557290915

456191LV00018B/791/P

9 781332 227822